Sustainable Energy Development

Environment and Society

Series Editor: Douglas Vakoch

As scholars examine the environmental challenges facing humanity, they increasingly recognize that solutions require a focus on the human causes and consequences of these threats, and not merely a focus on the scientific and technical issues. To meet this need, the Environment and Society series explores a broad range of topics in environmental studies from the perspectives of the social sciences and humanities. Books in this series help the reader understand contemporary environmental concerns, while offering concrete steps to address these problems.

Books in this series include both monographs and edited volumes that are grounded in the realities of ecological issues identified by the natural sciences. Our authors and contributors come from disciplines including but not limited to anthropology, architecture, area studies, communication studies, economics, ethics, gender studies, geography, history, law, pedagogy, philosophy, political science, psychology, religious studies, sociology, and theology. To foster a constructive dialogue between these researchers and environmental scientists, the Environment and Society series publishes work that is relevant to those engaged in environmental studies, while also being of interest to scholars from the author's primary discipline.

Recent Titles in the Series

Sustainable Energy Development: Technology and Investment, edited by Kimarie Engerman, Elizabeth A. Keenan, Dmitry Kurochkin, and Elena V. Shabliy
Making Nature Social: Towards a Relationship with Nature, by Rembrandt Zegers
Earth Polyphony: Law, Ecocriticism, and Eco-Activism in Postcolonial India, by Suhasini Vincent
Staying Together: Natureculture in a Changing World, Edited by Kaushani Mondal
Learning in the Anthropocene: Reimagining Education in the Twenty-First Century, by Carl A. Maida
The Social Life of Unsustainable Mass Consumption, by Magnus Boström
Everyday Life Ecologies: Sustainability, Crisis, Resistance, by Alice Dal Gobbo
Environmental Legacies of the Copernican Universe, by Jean-Marie Kauth
Anticipatory Environmental (Hi)Stories from Antiquity to the Anthropocene, edited by Christopher Schliephake and Evi Zemanek
Mapping the Environmental Humanities: The Emerging Role of GIS in Ecocriticism, edited by Mark Terry and Michael G. Hewson
The Bangladesh Environmental Humanities Reader: Environmental Justice, Developmental Victimhood, and Resistance, by Samina Luthfa, Mohammad Tanzimuddin Khan, and Munasir Kamal
Loren Eiseley's Writing across the Nature and Culture Divide, by Qianqian Cheng
The Saving Grave of America's Green Jeremiad, by John Gatta

Art and Nuclear Power: The Role of Culture in the Environmental Debate, by Anna Volkmar

Contesting Extinctions: Decolonial and Regenerative Futures, edited by Luis I. Prádanos, Ilaria Tabusso Marcyan, Suzanne McCullagh, and Catherine Wagner

Sustainable Energy Development

Technology and Investment

Edited by Kimarie Engerman, Elizabeth A. Keenan, Dmitry Kurochkin, and Elena V. Shabliy

LEXINGTON BOOKS
Lanham • Boulder • New York • London

Published by Lexington Books
An imprint of The Rowman & Littlefield Publishing Group, Inc.
4501 Forbes Boulevard, Suite 200, Lanham, Maryland 20706
www.rowman.com

86-90 Paul Street, London EC2A 4NE

British Library Cataloguing in Publication Information Available

Library of Congress Cataloging-in-Publication Data

Names: Engerman, Kimarie, editor. Title: Sustainable energy development :technology and investment / edited by Kimarie Engerman, Elizabeth A. Keenan, Dmitry Kurochkin, and Elena V. Shabliy. Description: Lanham : Lexington Books, 2024. | Series: Environment and society | Includes bibliographical references and index. | Summary: "This book provides deeper insights into sustainable energy development, technology, and investment in climate-oriented solutions; climate-aligned investment strategies present opportunities and challenges simultaneously"— Provided by publisher. Identifiers: LCCN 2024021238 (print) | LCCN 2024021239 (ebook) | ISBN 9781666965810 (cloth) | ISBN 9781666965827 (epub) | Subjects: LCSH: Sustainable development. | Sustainability. Classification: LCC HC79.E5 S86195 2024 (print) | LCC HC79.E5 (ebook) | DDC 338.9/27—dc23/eng/20240710 LC record available at https://lccn.loc.gov/2024021238LC ebook record available at https://lccn.loc.gov/2024021239

Contents

Introduction

International Climate Finance and Sustainably Responsible Investment

Elena V. Shabliy

Sustainable energy development is a critical step in addressing global problems faced by humanity, including human-induced climate change and the need for renewable energy development. Climate change is an urgent problem requiring an immediate reaction; innovative technologies and strategic investments should reshape the future energy landscape aimed at environmentally friendly and strategically valuable sustainable pathway. From a macroeconomic perspective, economic growth is impossible without innovative technologies, investments, and strategic planning. In March 2023, the United States' investment accounted for 20.7% of its Nominal GDP, with US clean energy investments comprising $141 billion.[1] The Intergovernmental Panel on Climate Change (IPCC) has emphasized the necessity of investing an average of US $2.4 trillion in clean energy every year in order to prevent climate change.[2]

The Biden-Harris administration considers climate change an *existential threat of our time*, prompting investors to integrate environmental practices in their day-to-day operations and decision-making processes.[3] The role of capitalism in future sustainable development depends on political discourses and opportunity structures. In particular, "green capitalism," valuing natural resources, could become an effective tool in climate change adaptation and mitigation, addressing both efficiency and equity.[4] China, with a socialist market economy, continues investing in renewable energy development, joining the EU and the US as significant global players. The US is the largest green bonds issuer, followed by China as the second-largest issuer.[5] Green bonds, representing debt instruments aimed at funding environmentally-oriented projects, can be issued by governments and corporations, alongside various financial institutions, to support projects with positive environmental outcomes.[6]

Chapter 1 of this book emphasizes that China has invested significantly in research and development (R&D); China worked on production of Critical Raw Materials (CRMs) that have a profound influence on renewable energy development. Government, venture capitalists, and companies allocate significant resources to support R&D, as well as the growth of green technology. Green bonds and impact investments attract capital towards environmentally-oriented projects. Developed economies plan to invest billions of dollars to tackle climate change, however, "the real cost is in the trillions," as Treasury Secretary Janet Yellen points out.[7] Nearly 200 countries approved a global climate pact that explicitly calls for eliminating fossil fuels like oil, gas and coal that are dangerously heating the planet.

The global shift to clean energy is inevitable; renewable energy is a fast-growing field with global renewable power capacity constituting almost two-thirds of total energy generation.[8] It has become competitive with fossil fuel. To further develop renewable technologies, financial, administrative, legal, and policy aspects should be considered and analyzed. Impactful policy design may help in changing old strategies to new strategies for combating climate change. From November 30 to December 3, 2023, the COP28 UN Climate Change Conference in Dubai was one of the largest climate-related world efforts, concluding the segment of addressing climate change under the Paris Agreement; around 85,000 participants represented civil society, businesses, national delegations, Indigenous Peoples, etc.[9] There was an urgent call on governments to accelerate the renewable energy growth and completely eliminate fossil fuels, considering slow progress in the recent sustainable transition.[10] The number of participant country parties proves the global commitment to climate change adaptation and mitigation strategies.

The understanding of this urgency is related to our realization of all the threats climate change brings to the global community. When comparing the current climate to the pre-industrial level, there is strong evidence that the global climate has changed, negatively impacting ecosystems, human well-being, and living organisms.[11] According to NASA's Goddard Institute for Space Studies (GISS), even a one-degree global change is significant—it takes a vast amount of heat to warm all of the oceans, the land mass, and the atmosphere by one degree.[12] Ocean species, marine life and habitats, and corals are in danger because of human-induced climate change.[13] It is obvious that climate is constantly changing, however, this time it happened due to human activity. Historical times mentioned by writers and scientists that is called Fertile Crescent, once considered the "Cradle of Civilization" (modern Israel, Iraq, Turkey, Syria, Lebanon, Palestine, perhaps Egypt), with rich agriculture, technological innovations, and writing.[14] The term

"Fertile Crescent," was first coined by University of Chicago Professor James Henry Breasted.[15] Notably, until the nineteenth century, scientists believed that ancient civilization began in Greece and Rome, and such scholars as Breasted, revolutionized this understanding with the idea of the ancient Middle East as the main origin of civilization.[16] Scholars believe that urbanization first took place in the *Fertile Crescent.*[17] *The Epic of Gilgamesh* is an ancient Mesopotamian epic poem and is believed to be one of the first literary works in human history.[18] In fact, *The Epic of Gilgamesh* gives an account of a great flood, a similar to the one in the Book of Genesis of the Old Testament. Researchers find that the climate during the time when the Fertile Crescent region supported ancient civilizations was different compared to the modern climate. The Fertile Crescent, and its changing climate, could encourage the evolution and spread of an "*r*-selected" type of annual plant species that produce edible seeds, in contrast to "*K*-selected" type of perennial plants.

Anthropogenic impact on climate change has been present for centuries; however, it began to accelerate in the mid-twentieth century. Since 1880, Earth's temperature has risen by an average 0.08ºC per decade (or about 2°F in total).[19] The rate of warming is more than twice as fast—0.18ºC—since 1981.[20] The ten warmest years in the historical record have occurred since 2010.[21] In collaboration with NASA, the Food and Agriculture Organization (FAO) measured temperature change for 1961–2021 at the global scale, demonstrating significant warming trends around the globe (The FAO, 2021).[22] Limiting temperature below 1.5ºC is essential for the global well-being and current complex ecosystems. The year 2023 was recorded as the warmest, with the global mean temperature change surpassing the climate normal—1.7ºC.[23] The most significant temperature changes were recorded in Europe, Asia, Oceania, South America, North America, and Africa.[24] In 2020, the largest mean annual temperature change occurred in Europe, followed by other regions. Additionally, existing data for global natural disasters demonstrate a tendency toward an increasing number of extreme weather events.[25] For example, in 2011–2012, the United States saw fourteen weather disasters, each causing damages exceeding $1 billion. These disasters included tornadoes, hurricanes, wildfires, flooding, drought, etc.[26] As the GISS finds, global warming does not mean temperatures rise everywhere at every time by same rate.[27] The temperatures we experience locally fluctuate due to many factors, such as, for example, wind and precipitation patterns.[28] The global temperature depends on how much energy is received from the Sun and how much the Earth radiates this energy back to space.[29] The amount of energy received from the Sun does not change radically every year, energy radiated by Earth is closely related to the chemical composition of the atmosphere, such as greenhouse gasses (GHGs).[30]

The historical temperature records indicate that temperature changes occur globally in various countries and territories. The problem of temperature increases is likely caused by the Industrial Revolution and the GHGs. The Linear Regression Model we present below shows statistical evidence of temperature increase, with a 95% confidence interval for the rate of temperature change (in Celsius) per year being (0.023, 0.028). To demonstrate global climate change, we run a statistical Linear Regression Model representing global temperature change in 192 countries and 38 territories. The dataset consists of 13,764 values of Temperature Change (with respect to a baseline climatology) at 247 locations measured over the period of 1961–2021. This data was retrieved from the Food and Agriculture Organization (https://www.fao.org/faostat/en/#data/ET) in 2023. This source presents statistics of land surface air temperature, showing annual updates starting from 1961 and including information up to 2021. The standard deviation (of the temperature change) of the methodology is also present. For earlier periods (starting from 1880), the data can be obtained from the National Aeronautics and Space Administration Goddard Institute for Space Studies (NASA-GISS).[31] In 2021, statistics demonstrated significant warming trends worldwide; approximately 170 countries and territories recorded higher-than-usual mean annual temperatures.[32]

From Figure 0.1, we can clearly see a positive trend of temperature change over 1961–2021. For a given year, we may have multiple observations at neighboring locations which are potentially highly correlated. To mitigate the issue of non-independence of errors in the linear regression, we first aggregate data for each year, so that the response variable is the Average Temperature Change over all locations for a given year. Results together with the trendline and 95% confidence intervals for expected Average Temperature Change (for a given year) are shown in Figure 0.2 below. The slope of the Linear Model shows a warming trend, explaining a substantial portion of variance.

Here, we model the dependence of Average Temperature Change on year using the Linear Regression Model.

$$\text{Average Temperature Changei}=\beta_0+\beta_1\ \text{year}_i+\text{error}_i.$$

Here, index *i* enumerates aggregated observations, that it runs from 1 to 61. Using Ordinary Least Squares (OLS), the estimates of β_0 and β_1 in this model are given by β_0 = -50.52 and β_1= 0.026, respectively. We can see that the slope is estimated to be positive, and temperature rises approximately at the rate of 2.6 ºC per 10 years. P-value for the test H_0: β_1=0 vs. H_1: β_1>0 is less than $1\cdot10^{-16}$ (this p-value is obtained as the p-value of the two-sided test divided by 2—please see the regression output below). At level α= 0.05, there is evidence that β_1>0. The 95% confidence interval for β_1 is (0.023, 0.028).

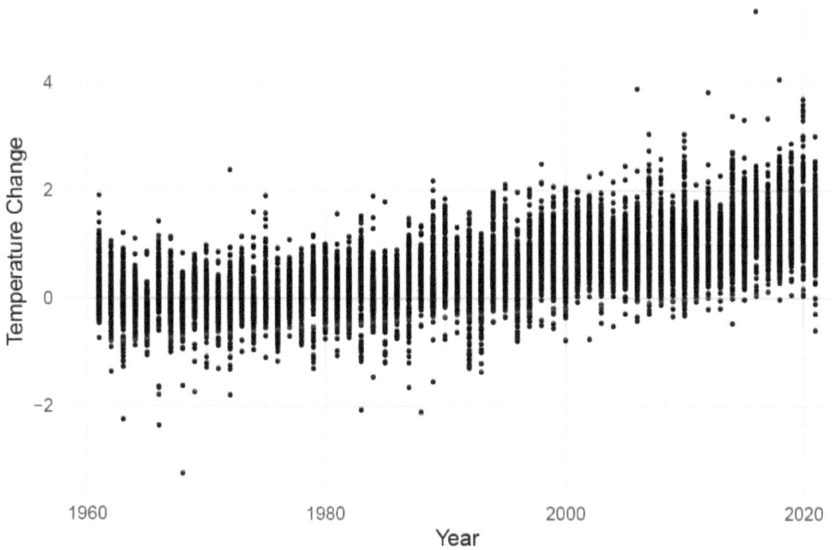

Figure 0.1. Temperature Change.
Source: Created by the author of this chapter.

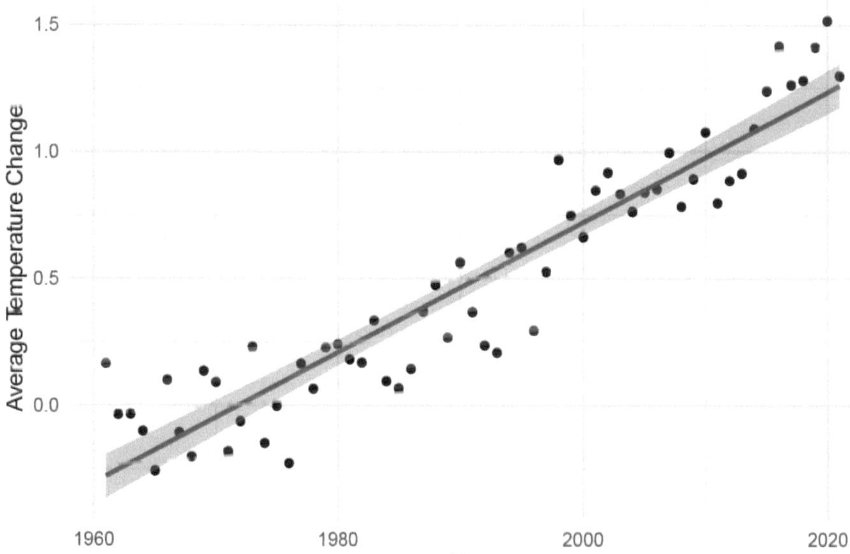

Figure 0.2. Dependence of Average Temperature Change (with respect to a baseline climatology) on Year. Blue line is the trendline. Ninety-five percent confidence intervals for expected Average Temperature Change (for a given year) are shown in grey.
Source: Created by the author of this chapter.

Below is the summary for the OLS regression:

```
Residuals:
Min 1Q Median 3Q Max
-0.33311 -0.11691 0.01765 0.10055 0.44382

Coefficients:
Estimate Std. Error t value Pr(>|t|)
(Intercept) -50.518615 2.472008–20.44 <2e-16 ***
Year 0.025620 0.001242 20.64 <2e-16 ***
---
Signif. codes: 0 '***' 0.001 '**' 0.01 '*' 0.05 '.' 0.1 ' ' 1

Residual standard error: 0.1707 on 59 degrees of freedom
Multiple R-squared: 0.8783,     Adjusted R-squared: 0.8762
F-statistic: 425.8 on 1 and 59 DF, p-value: < 2.2e-16
```

We see that 88% of variance of the Average Temperature Change can be explained by the year variable. We can also look at the distribution of residuals to evaluate if they can be assumed to be coming from Normal distribution. Below is Figure 0.3 that displays the residual distribution (left chart) with added Normal curve (indicated in red) and the Normal quantile-quantile plot (right chart). From the distribution of the residuals (Figure 0.3, left) one can see that the residuals are close to Normal but still could be different. From the Normal quantile-quantile plot (Figure 0.3, right) one can see that the distribution of residuals does not appear to deviate from Normal. We first run Shapiro-Wilk test for normality:

```
Shapiro-Wilk normality test


W = 0.98894, p-value = 0.8574
```

and observe that there is indeed no evidence (at level α=0.05) that the distribution of residuals is different from Normal. However, if we run the Kolmogorov-Smirnov test to compare the distribution against normal:

```
One-sample Kolmogorov-Smirnov test


D = 0.36953, p-value = 5.572e-08
alternative hypothesis: two-sided
```

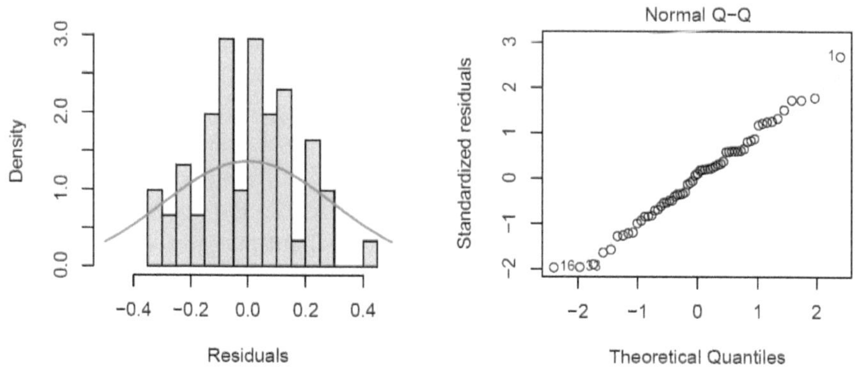

Figure 0.3. Distribution of Linear Model Observed Residuals (left) with Normal Curve Added (in red) and the Normal Q-Q Plot of the Residuals (right chart).

Source: Created by the author of this chapter.

We finally see evidence (at level α=0.05) that the distribution of residuals is different from Normal.

Nevertheless, the sample size has 61 observations so we still can use the obtained p-value and the 95% confidence interval for β_1 because the sample is large enough to use the Normal approximation to the distribution of β_1. There is clear evidence of a long-term positive trend in global average temperature change; the model presents clear statistical evidence of a long-term positive trend in global average temperature change in the twentieth and twenty-first centuries.

THE INVESTOR AGENDA

Although the global pandemic and current conflicts in the world have had a significant impact on the climate change agenda, the allocation of capital into net zero emission solutions and financing environmentally-friendly initiatives has remained strong.[33] The Investor Agenda is a collaborative initiative among a group of international investors that launched in San Francisco in 2018; the investors are committed to climate change dialogue, promoting sustainability in their investment strategies and practices. The Investor Agenda represents a type of leadership action on the climate crisis that is also comprehensive and focused on investor mobilization for a net-zero emissions economy.[34] The Investor Agenda has produced a new comprehensive framework for Investor Climate Action Plans (ICAPs).[35] The Agenda suggests new investing strategies focusing on the elimination of thermal coal investment, as coal is one of the major contributors to global GHG emissions; and climate change has been integrated into the long-term investment strategies and policy advancement.[36] The Asia Investor Group on Climate Change calls for action among Asia's financial organizations, emphasizing financial opportunities and challenges of green investing. The Investor Group on Climate Change (IGCC) brings together investors with total funds under management of over $2 trillion in Australia and New Zealand and $20 trillion around the world. IGCC members cover over 7.5 million people in Australia and New Zealand.[37] Finance plays an important role in the acceleration of the clean energy shift and building of a climate-resilient perspective. The existing financial resources are insufficient to tackle the climate crisis; President Biden's global climate finance plan aims to address this gap, supporting developing countries and encouraging the emphasis on investment aligned with climate goals. To more effectively mobilize public and private finance, the United States announced it is:

1. Scaling up international financing to address climate needs.
2. Issuing the first U.S. International Climate Finance Plan.
3. Launching an international dialogue on decreasing fiscal climate risk through national budgets.[38]

The US aims to develop green technologies.[39] In addition, the US establishes a U.S.-India Climate and Clean Energy Agenda 2030 Partnership. The United States is working with partners around the world to "set ambitious 2030 targets for climate action and clean energy innovation and deployment."[40] As the White House website states, "The U.S.-India Climate and Clean Energy Agenda 2030 Partnership will elevate ambitious climate action as a core theme of US-India collaboration and support the achievement of India's ambitious targets, including reaching 450 GW of renewable energy by 2030."[41] The US economic model is based on freedom, liberty, and other principles of democracy—it raises standards of living and promotes competition, entrepreneurship, innovation, and technology.[42] The free-market system is the best way for generating new jobs in sustainability, a strong economy, innovation and advanced technology, as well as an economic opportunity for all.[43]

This book provides deeper insights into sustainable energy development, technology, and investment in climate-oriented solutions; climate-aligned investment strategies present opportunities and challenges simultaneously. The interrelation of the environment and the green economy is emphasized in this book. Using appropriate financial instruments will be crucial in the successful trajectory of sustainable pathways. Finance presupposes that people are risk-averse; however, "[k]nowing that people will trade off risk for return and that most people are risk averse, we can start to understand a lot of fundamental financial concepts, including risk management tools, hedging strategies, forward or future contracts, the equity risk premium, insurance policy pricing."[44] Gregory L. Sabin explains that finance is "about earning a higher return (or yield) on an investment than the cost of capital that was used to fund the project, considering all available alternative projects, appetites for risk, and differences in the time in the timing and riskiness of cash flow."[45] Businesses are foundational for a strong economy and creating jobs; they promote innovation and technology. Good corporate governance with a sustainable emphasis can attract investments and benefit other stakeholders. The main five principles of Corporate Governance are awareness and accountability, transparency and fairness—and last but not least—*responsibility*. More and more companies are becoming more responsible, investing their efforts in protecting the environment by adopting more sustainable practices.

Environmental, Social, and Corporate Governance (ESG) are sound criteria that investors consider when assessing companies' impact on the environment and its social practices, as well as corporate governance structure, and it has gained popularity among stakeholders, including investors, with a strong emphasis on sustainability. The ESG policies are incorporated in the companies' long-term strategies.[46] The green bond market has been developing rapidly; green bonds are becoming powerful financial tools aligning with sustainable practices and pathways, presenting sustainable finance development. However, when issuing green bonds some challenges may occur. For example, identifying a green criterion is a challenge. Here, aspects such as adherence to green standards, transparency, and accountability, come into play. These aspects are important in presenting trustworthy green projects. Notably, most businesses tend to comply with green standards, as this is an essential part of sustainable pathway. Sometimes limited liquidity also may be less attractive for investors when they buy or sell green bonds. Another challenge associated with green bonds is uncertainties regarding regulations that could affect economic efficiency. Additional costs may be involved when there is a requirement to assess green projects. Despite these challenges, global commitment to climate change prevention makes the green bond market more attractive to investors. The Climate Bonds Initiative plays a salient role in the development and growth of the green bond market. The European Union's Green Taxonomy provides a clear Green Bond Framework facilitating the development of green bond market. Sustainable Finance is associated with SRI (Socially or Sustainably Responsible Investment), ESG (Environmental, Social, and Corporate Governance), and CSR (Corporate Social Responsibility).[47] The European Commission provides the following definition of sustainable finance: "sustainable finance is understood as finance to support economic growth while reducing pressures on the environment to help reach the climate- and environmental objectives of the European Green Deal, taking into account social and governance aspects."[48] The key idea is that "finance should make an appropriate contribution to sustainability."[49] Long-term orientation is one of the main features of sustainable finance.[50]

This edited collection emphasizes the urgency of the clean energy transition, presenting chapters discussing investment and the development of renewable energy technology. A group of 631 investors from around the world signed a letter in 2019, calling on governments to concentrate their efforts on climate change and eventually achieve the goals of the Paris Agreement. In an open letter to the Governments of World, the CEOs of the seven partner organizations of The Investor Agenda supported the IPCC (Intergovernmental Panel on Climate Change) special report on 1.5°C and reiterated to world governments the urgency of achieving the goals set by the Paris Agreement.

In the recent past, many investors and companies gradually altered their strategies, focusing their efforts on sustainable trends. More companies have become engaged and interested in research related to the economic dimensions of climate change. Reducing energy consumption and cutting of GHGs are the leading ways that could contribute to the objectives and targets set by the IPCC. Cutting GHGs through the advancement of climate change policy can be seen as a valuable component of the broader sustainability movement. Renewable Energy Sources (RES) have gained public acceptance worldwide and demonstrated impressive progress.

The International Energy Agency (IEA) points out that innovations in energy technology can bring more benefits and facilitate transformation, but advancing policy is integral. Undoubtedly, the COVID-19 crisis influenced the energy sector and future investment trends; recovering from the COVID-19 crisis may impact sectors differently. For example, low prices and low demand in all subsectors will leave energy companies with weakened financial outcomes. The IPCC advises investing US $2.4 trillion in clean energy yearly through 2035 to prevent a global temperature rise.

Chapter 1 suggests that governments can deploy tax incentives, targeted subsidies to foster the development of new technologies, provide favorable lending, and promote sustainable finance. Critical Raw Materials (CRMs) have emerged as one of the most highlighted topics in academic research and environmental policy decision-making over the last decade. As countries continue to demand more of such critical materials, the authors argue that this topic is expected to play a key role across the globe in the next decades, not only in climate change mitigation or green transition issues but also in many areas from supply chain management to modern technological progress; it is now considered that this multifaceted topic significantly affects the current material demand trends, posing risks to be addressed and requiring prioritization in environmental policies. This chapter seeks to understand the crucial role of CRMs in the clean energy transition.

In Chapter 2, Skoglund concentrates the Swedish steel industry, arguing that energy transitions have continuously impacted steel production processes. This chapter traces such future-making in the conflation of industrial and climate policy expectations within the Swedish steel industry. As a complement to existing knowledge about rational-decision making, the chapter explains how metaphors and expectations also contribute to the realization of transformations. In Chapter 3, Gordon seeks to advance the international environmental justice debate by positing that domestic environmental policy terms of *best available* and *best control technology* the Federal Water Pollution Control Act warrant closer evaluation and could be useful measuring approaches in creating a positive impact for mitigating international environmental justice issues. Thid chapter emphasizes the importance of environmental protection, supporting *intergenerational* environmental justice and reminding existing vulnerable groups that will be directly impacted by climate change. This chapter also discusses the interconnection of environmental justice, sustainable development, technology, and investment. It underscores the need for investment in sustainable pathways and innovative, technology-oriented approaches to combat climate change. Chapter 4 focuses on the moving the state Michigan towards just building; it is comprised of participant observation of the Michigan Environment, Great Lakes, and Energy (EGLE) Council on Climate Solutions open working subgroups of (1) building and housing and (2) energy production, distribution and storage. In Chapter 5, Nascimento addresses climate change in Bahia state, Brazil, and Brazilian development policy and natural resources of the country; hydropower, she argues, was the major type of electric energy for nearly fifty years, but wind power generation has become a new trend in the last decades. The recent growth of energy generation in the state still takes place due to the increase in renewable energy, but the new power now comes mainly from the wind and is integrated into the centralized national electricity generation and transmission system. This work contributes to the literature on policy, renewable energy development in Brazil and Bahia, environmental and socio-economic studies. Policy and the natural endowment of the state continue to influence the expansion of electric energy production in Bahia and the diversification of sources.

Chapter 6 concentrates on technology and investment aspects, proposing a novel framework that integrates top-down and bottom-up grid management paradigms to enable the decentralization of distribution systems market operations; this framework leverages existing smart grid technologies such as Virtual Power Plants, Microgrids, and Distributed Energy Resources, and redefines their roles to align with the bottom-up approach that empowers prosumers. It also ensures compatibility with the top-down grid operations mechanisms by coordinating the operational timescales of different levels, adopting a hierarchical design that consists of three levels: a *peer-to-peer* energy market level, a *microgrid* level, and a *virtual power plant* level. The *peer-to-peer* energy market level is the lowest level where prosumers interact with each other to trade energy locally. The *microgrid* level is the middle level where microgrids aggregate prosumers and act as intermediaries between them and the *virtual power plants*. The *virtual power plant* level is the highest level where virtual power plants coordinate the energy supply and demand of multiple microgrids and interface with the wholesale market. The chapter demonstrates how the hierarchical framework can decentralize the distribution systems market operations using a bottom-up perspective while addressing the challenges of scalability and interoperability.

Chapter 7 emphasizes that the literature on clean energy transition and related policies often focus on technological solutions and innovation. Gz. MeeNilankco Theiventhran argues that technology has significantly influenced and shaped human life, with the expectation that science and technology can address global challenges, such as climate change. Energy transitions primarily concentrate on renewable technologies, where innovations will create pathways for the widespread diffusion of renewables even in unexpected places, as illustrated by the case of rooftop solar uptake in Sri Lanka. Gz. MeeNilankco Theiventhran points out that this case contributes to understanding critical questions and highlights essential aspects of energy transitions in the global South, showing how interactions between policy, finance, and technology can leave certain groups of people behind. Renewable projects create new inequities and deepen existing ones, illustrating how Sri Lanka's post-war political context has created peripheries within the periphery, as the author put it.

Chapter 8 discusses the Indian success in the sustainable development pathway. India is one of rapidly growing economies that achieved its independence in 1947. Notably, it is the second most populous country in the world. With its rising population, India's demand for energy is also increasing. According to the IEA, India is the world's third-largest energy-consuming country, due to rising incomes and improving standards of living. At the time of its independence, India heavily relied on coal for meeting its energy demands, leading to more greenhouse gas emissions (GHG) emissions in the atmosphere. Since then, it has been exploring alternative sources of energy in alignment with the Paris Agreement. India has opened multiple investment opportunities and engaged in multilateral partnerships globally, highlighting the strategic importance of renewable energy in Indian foreign policy. Presently, India stands the world's third largest producer of renewable energy. The chapter presents a case study analysis of India's multilateralism in renewable energy, using an inductive method. Chapter 9 elaborates on the importance of climate change education in Nigeria. The authors conclude that climate change education is a necessity and a mandatory practice to effectively respond to the environmental needs of the contemporary world especially in developing countries like Nigeria. Moreover, education is a form of human capital, and investing in climate change education is an important step for tackling this urgent issue.

THE BUSINESS ROUNDTABLE AND SUSTAINABLE DEVELOPMENT

We cannot underestimate the role of businesses in the sustainable pathway. The Business Roundtable has updated its principles regarding the role of a corporation by issuing Principles of Corporate Governance, which include the purpose of a corporation. These Principles have evolved over time, having been issued since 1978. The 2019 statement presents a shift with a new approach—serving not only shareholders but also communities, customers, employees, suppliers, in other words, all *stakeholders*. The main principles of Corporate Governance are awareness and accountability, transparency and impartiality, and last but not least, social *responsibility*. Corporate Governance principles also include a broader range of other aspects, in addition to those mentioned above, creating a system of rules and practices that determines how a company operates and how it aligns with the interest of *all its stakeholders*. Abandoning shareholder primacy was the primary shift, emphasizing service to *all stakeholders*. Notably, good corporate governance can attract more investors because it also proves ethical business practices commitment, such as sustainability. One of the challenges of this approach is fair treatment; each *stakeholder*, through hard work, should have a successful pathway. Fair treatment for *all stakeholders* involves such aspects as the equitable distribution of benefits, opportunities, and responsibilities. In other words, all the theoretical underpinnings should be reflected in day-to-day operations. Business leaders should demonstrate better corporate principles, fully embracing this new paradigm that is considered one of the best business paradigms, advancing gender equity and equality, as well as sustainability; promoting the sustainable development that also includes fair treatment is one the most important aspects of corporate social responsibility. Ensuring fair treatment for *all stakeholders* involves addressing issues such as the equitable distribution of benefits, opportunities, and responsibilities. This modernization will propel businesses forward, aligning with environmental awareness and promotion of sustainable practices, as the Business Roundtable recognizes sustainability as an essential aspect for future business success.

CONCLUSION

Over three decades ago, the severity of environmental problems have been recognized and acknowledged scientifically.[51] The topic of climate change has been of great importance since then, and keeping temperature under 1.5ºC, a level that researcher and scientists consider as a "dangerous threshold," is essential for a global well-being; not exceeding 1.5ºC will require reductions in carbon dioxide emissions.[52] The sustainable revolution of the energy sector requires investments in clean energy technology and innovations. Climate change and other related problems continue to harm the global community and serve as the impetus for advancing the green energy economy worldwide, along with the global energy justice movement. Research and Development (R&D), as well as investment into sustainability, will undoubtedly contribute to climate change mitigation and adaptation, fostering economic development and stimulating the creation of new jobs in sustainability and RES. The role of banks regarding the sustainable development is significant "considering the intermediary role that they play in society."[53] There is a growing body of literature on the subject of climate change; however, there is an urgent need for further investigation of the problem of global temperature increase and climate change (Shabliy et al., 2021). According to the IPCC Climate Change 2023: Synthesis Report, it is necessary and vital to keep the global temperature increase below 1.5°C versus higher levels.[54] The statistical Linear Regression Model presented above proves that temperature change is evident at the global scale throughout 60 years. We can see that the slope is estimated to be positive, and temperature rises approximately at the rate of 2.6ºC per 10 years. Climate change may cause more disasters on the planet; climate scientists expect that the hydrologic cycle (the continuous movement of water on, above and below the surface) will intensify in a warming climate.[55] As a consequence, a warmer climate will increase evaporation and precipitation rate globally. The present temperature change statistics "document recent warming trends in all countries in the world, facilitating public understanding of the climate change challenges to agriculture," as well as "helping to identify possible responses necessary to minimize risk to food production."[56] It is important to point out that extra heat is driving regional and seasonal temperature extremes; it reduces snow cover and sea ice; it causes heavy rainfall; it changes habitat ranges for plants and animals.[57] It is important to limit global warming under 1.5° Celsius above pre-industrial levels, as it is determined by the Paris Agreement. Global GHGs will peak before 2025. The use of coal should be reduced to almost zero by 2050, according to the IPCC. GHGs should decline by more than 40% by 2030 and net-zero by 2050.[58] Many countries are already developing *climate action* plans aimed at cutting GHGs and adapting to climate change devastating impacts.

NOTES

1. "United States Investment: % of GDP" https://www.ceicdata.com/en/indicator/united-states/investment--nominal-gdp (last accessed 12/29/2023); https://www.statista.com/statistics/499193/clean-energy-investment-in-the-us/ (last accessed 2/20/2024).
2. "FACT SHEET: Biden-Harris Administration Leverages Historic U.S. Climate Leadership at Home and Abroad to Urge Countries to Accelerate Global Climate Action at U.N. Climate Conference (COP28)," https://www.whitehouse.gov/briefing-room/statements-releases/2023/12/02/fact-sheet biden-harris-administration-leverages-historic-u-s-climate-leadership-at-home-and-abroad-to-urge-countries-to-accelerate-global-climate-action-at-u-n-climate-conference-cop (last accessed 2/27/2024).
3. Ibid.
4. Graciela Chichilnisky (2019). "Green Capitalism." *Journal of International Affairs* (New York), 73(1), 161–170.
5. "China's Green-Bond Market: Growing Issuance and Historical Outperformance," https://www.msci.com/www/blog-posts/china-s-green-bond-market/03029315810 (last accessed 2/20/2024).
6. In terms of structure, green bonds share similar characteristics as traditional bonds; however, green bonds have specific characteristics ensuring accountability and transparency in the use of proceeds.
7. "Janet Yellen, the Treasury Secretary, Says the U.S. will Help Raise $500 Million for Green Bonds," https://www.nytimes.com/2021/11/03/world/yellen-green-bonds-climate-change.html (last accessed 2/20/2024).
8. Kurochkin et al. (2019). *Renewable Energy: International Perspectives on Sustainability* (1st ed.). Palgrave Macmillan.
9. "UN Climate Change Conference—United Arab Emirates," https://unfccc.int/cop28 (last accessed 2/27/204).
10. Ibid.
11. "Impacts of 1.5°C Global Warming on Natural and Human Systems" https://www.ipcc.ch/sr15/chapter/chapter-3/ (last accessed 4/1/2021).
12. "World of Change: Global Temperatures," https://earthobservatory.nasa.gov/world-of-change/global-temperatures (last accessed 4/8/2023).
13. "Climate Change Driving Entire Planet to Dangerous 'Tipping Point,'" https://www.nationalgeographic.com/science/article/earth-tipping-point (last accessed 2/27/2024).
14. "Fertile Crescent," https://www.history.com/topics/pre-history/fertile-crescent (last accessed 1/16/2024).
15. Ibid.

16. "The Fertile Crescent, Explained," https://news.uchicago.edu/explainer/fertile-crescent-explained (last accessed 1/16/2024).

17. Ibid.

18. Ibid.

19. "Climate Change: Global Temperature," https://www.climate.gov/news-features/understanding-climate/climate-change-global-temperature#:~:text=Earth's%20temperature%20has%20risen%20by,0.18%C2%B0%20C)%20per%20decade (last accessed 4/8/2023).

20. "Impacts of 1.5°C Global Warming on Natural and Human Systems" https://www.ipcc.ch/sr15/chapter/chapter-3/ (last accessed 4/4/2021).

21. "Climate Change: Global Temperature," https://www.climate.gov/news-features/understanding-climate/climate-change-global-temperature#:~:text=Earth's%20temperature%20has%20risen%20by,0.18%C2%B0%20C)%20per%20decade. (last accessed 4/8/2023).

22. "Temperature Change Statistics 1961–2020 Global, Regional and Country Trends," https://www.fao.org/3/cb4410en/cb4410en.pdf (last accessed 4/4/2021). These statistics covered 192 countries as well as 38 territories.

23. Ibid.; "NASA Analysis Confirms 2023 as Warmest Year on Record," https://www.nasa.gov/news-release/nasa-analysis-confirms-2023-as-warmest-year-on-record/ (last accessed 2/20/2024).

24. Ibid.

25. J. F. C. DiMento& P. Doughman. (2014). *Climate Change: What It Means for Us, Our Children, and Our Grandchildren* (2nd edition). MIT Press.

26. Ibid.

27. "World of Change: Global Temperatures," https://earthobservatory.nasa.gov/world-of-change/global-temperatures (last accessed 4/8/2023).

28. Ibid.

29. Ibid.

30. Ibid.

31. "GISS Surface Temperature Analysis (GISTEMP v4)," https://data.giss.nasa.gov/gistemp/ (last accessed 4/21/2023).

32. "Temperature Change Statistics 1961–2021. Global, Regional and Country Trends," https://www.fao.org/food-agriculture-statistics/data-release/data-release-detail/fr/c/1492093/ (last accessed 4/21/2023).

33. "2020 Net Zero Investment Survey," https://igcc.org.au/wp-content/uploads/2020/10/Oct2020_Final-IGCC-Net-Zero-Investment-Report.pdf (last accessed 1/16/2024).

34. https://theinvestoragenda.org (last accessed 1/16/2024).

35. Ibid.

36. Ibid.

37. "Accelerating Investment for a Net Zero Economy," https://igcc.org.au (last accessed 3/8/2024).

38. "FACT SHEET: President Biden's Leaders Summit on Climate," https://www.whitehouse.gov/briefing-room/statements-releases/2021/04/23/fact-sheet-president-bidens-leaders-summit-on-climate/ (last accessed 3/8/2024).

39. Ibid.
40. Ibid.
41. Ibid.
42. "Our Commitment," https://opportunity.businessroundtable.org/ourcommitment/ (last accessed 1/23/2024).
43. https://opportunity.businessroundtable.org/ourcommitment/ (last accessed 1/23/2024).
44. Gregory L. Sabin. *Principles of Finance in 15 Days.* Independently Published, (September 6, 2022), p. 12.
45. Ibid., p. 10.
46. U.S. Bhutta, A. Tariq, M. Farrukh, A. Raza, & M.K. Iqbal. (2022). "Green Bonds for Sustainable Development: Review of Literature on Development and Impact of Green Bonds." *Technological Forecasting & Social Change*, 175, 121378. https://doi.org/10.1016/j.techfore.2021.121378, p. 1.
47. L.T. Starks. (2023). "Presidential Address: Sustainable Finance and ESG Issues—Value versus Values." *The Journal of Finance* (New York), 78(4), 1837–1872. https://doi.org/10.1111/jofi.13255
48. "What Is Sustainable Finance?" https://finance.ec.europa.eu/sustainable-finance/overview-sustainable-finance_en#:~:text=In%20the%20EU%27s%20policy%20context%2C%20sustainable%20finance%20is,Deal%2C%20taking%20into%20account%20social%20and%20governance%20aspects (last accessed 2/27/2024).
49. P.K. Ozili. (2021). "Making Sustainable Finance Sustainable." *Financial Internet Quarterly*, 17(3), 65. https://doi.org/10.2478/fiqf-2021-0020 (last accessed 2/27/2024).
50. Ibid., 67.
51. Marcel Jeucken. (2010). *Sustainable Finance and Banking: The Financial Sector and the Future of the Planet* (1st edition). Routledge, an imprint of Taylor and Francis.
52.
53. Ibid.
54. "Impacts of 1.5ºC Global Warming on Natural and Human Systems," https://www.ipcc.ch/sr15/chapter/chapter-3/ (last accessed 4/4/2021).
55. J. F. C. DiMento & P. Doughman. (2014). *Climate Change: What It Means for Us, Our Children, and Our Grandchildren* (2nd edition). MIT Press.
56. "Temperature Change Statistics 1961–2020 Global, Regional and Country Trends," https://www.fao.org/3/cb4410en/cb4410en.pdf (last accessed 4/8/2021).
57. "Climate Change: Global Temperature," https://www.climate.gov/news-features/understanding-climate/climate-change-global-temperature#:~:text=Earth's%20temperature%20has%20risen%20by,0.18%C2%B0%20C)%20per%20decade (last accessed 4/8/2023).
58. "Goal 13: Take Urgent Action to Combat Climate Change and Its Impacts."

BIBLIOGRAPHY

Chichilnisky, G. (2019). "Green Capitalism." *Journal of International Affairs* (New York), 73(1), 161–170.

"China's Green-Bond Market: Growing Issuance and Historical Outperformance," https://www.msci.com/www/blog-posts/china-s-green-bond-market/03029315810 (last accessed 2/20/2024).

"Climate Change Driving Entire Planet to Dangerous 'Tipping Point,'" https://www.nationalgeographic.com/science/article/earth-tipping-point (last accessed 2/27/2024).

"Climate Change: Global Temperature," https://www.climate.gov/news-features/understanding-climate/climate-change-global-temperature#:~:text=Earth's%20temperature%20has%20risen%20by,0.18%C2%B0%20C)%20per%20decade (last accessed 4/8/2023).

DiMento, J. F. C., & Doughman, P. (2014). *Climate Change: What It Means for Us, Our Children, and Our Grandchildren* (2nd edition). MIT Press.

"FACT SHEET: Biden-Harris Administration Leverages Historic U.S. Climate Leadership at Home and Abroad to Urge Countries to Accelerate Global Climate Action at U.N. Climate Conference (COP28)," https://www.whitehouse.gov/briefing-room/statements-releases/2023/12/02/fact-sheet-biden-harris-administration-leverages-historic-u-s-climate-leadership-at-home-and-abroad-to-urge-countries-to-accelerate-global-climate-action-at-u-n-climate-conference-cop (last accessed 2/27/2024).

"Fertile Crescent," https://www.history.com/topics/pre-history/fertile-crescent (last accessed 1/16/2024).

GISS Surface Temperature Analysis (GISTEMP v4), https://data.giss.nasa.gov/gistemp/

"Goal 13: Take Urgent Action to Combat Climate Change and Its Impacts," https://www.un.org/sustainabledevelopment/climate-change/#:~:text=The%20global%20annual%20mean%20temperature,of%20the%20next%20five%20years (last accessed 4/21/2023).

"Impacts of 1.5ºC Global Warming on Natural and Human Systems," https://www.ipcc.ch/sr15/chapter/chapter-3/ (last accessed 4/4/2021).

"Janet Yellen, the Treasury Secretary, Says the U.S. will Help Raise $500 Million for Green Bonds," https://www.nytimes.com/2021/11/03/world/yellen-green-bonds-climate-change.html (last accessed 2/20/2024).

Jeucken, M. (2010). *Sustainable Finance and Banking: The Financial Sector and the Future of the Planet* (1st edition). Routledge, an imprint of Taylor and Francis.

"Our Commitment," https://opportunity.businessroundtable.org/ourcommitment/ (last accessed 1/23/2024).

Ozili, P. K. (2021). "Making Sustainable Finance Sustainable." *Financial Internet Quarterly*, 17(3), 64–70. https://doi.org/10.2478/fiqf-2021–0020

Sabin, Gregory L. *Principles of Finance in 15 Days.* Independently Published (September 6, 2022).

Shabliy E.V., Crawford M., and Kurochkin D., (eds.). *Energy Justice: Climate Change Mitigation and Adaptation*. Palgrave Macmillan, 2022.

Starks, L.T. (2023). "Presidential Address: Sustainable Finance and ESG Issues—Value versus Values." *The Journal of Finance* (New York), 78(4), 1837–1872. https://doi.org/10.1111/jofi.13255

"Temperature Change Statistics 1961–2020 Global, Regional and Country Trends," https://www.fao.org/3/cb4410en/cb4410en.pdf (last accessed 4/4/2021).

"Temperature Change Statistics 1961–2021. Global, Regional and Country Trends," https://www.fao.org/food-agriculture-statistics/data-release/data-release-detail/fr/c/1492093/ (last accessed 4/21/2023).

"United States Investment: % of GDP," https://www.ceicdata.com/en/indicator/united-states/investment--nominal-gdp (last accessed 12/29/2023).

"What Is Sustainable Finance?" https://finance.ec.europa.eu/sustainable-finance/overview-sustainable-finance_en#:~:text=In%20the%20EU%27s%20policy%20context%2C%20sustainable%20finance%20is,Deal%2C%20taking%20into%20account%20social%20and%20governance%20aspects (last accessed 2/27/2024).

"World of Change: Global Temperatures," https://earthobservatory.nasa.gov/world-of-change/global-temperatures (last accessed 4/8/2023).

Chapter One

Critical Raw Materials in the Clean Energy Transition

Evaluating Circular Economy and Investment Strategies for Mitigating Supply Risk

Etem Karakaya, Sedat Alataş, and Burcu Hiçyılmaz

At the 26th United Nations Climate Change Conference of the Parties (COP26) in Glasgow in late 2021, 197 parties agreed to the Glasgow Climate Pact. Accordingly, over 90% of the world's greenhouse gas (GHG) emissions are now covered by net-zero commitments. Besides, 153 countries agreed to revisit and strengthen their current emissions reduction targets, known as Nationally Determined Contributions (NDCs). These net-zero and emissions reduction targets clearly reveal that countries are highly ambitious to accelerate the global energy transition toward net-zero and to keep the goal of limiting temperature rise to 2°C (preferably to 1.5°C). However, as also stated in the Glasgow Climate Pact, countries' commitments alone are not sufficient to be successful. The achievement of these ambitious climate goals largely depends on countries' immediate implementation of climate actions on many issues aligning with their net-zero and 2030 targets, such as resource efficiency, renewable energy, and environmental technologies (United Nations, 2021). In this regard, some recent reports (European Commission, 2020a; International Energy Agency, 2022a; United Nations Environment Programme International Resource Panel, 2020) and papers (Ali et al., 2017; Girtan et al., 2021; Hertwich et al., 2015; Pommeret et al., 2022; Schandl et al., 2016) highlight the growing importance of the raw materials and clean energy technologies using these raw materials intensively.

Raw materials are considered economically vital for some key industries as they are used as an input for producing a wide range of goods and services, such as tungsten for smartphones, gallium for many light-emitting diodes (LEDs), copper for electronics, and silicon metal for semiconductors. Therefore, they are closely linked to our daily lives and essential parts of modern society (Ali et al., 2017; European Commission, 2020a; Kastanaki & Giannis, 2022). On the other hand, this is not the only reason why we pay special attention to them. Raw materials are also crucial for the transition toward a green economy and tackling climate change. It is because the achievement of net-zero goals requires advancing low-carbon technologies that employ various raw materials in substantial amounts, such as renewable energy (solar panels, wind turbines), electric vehicles (EVs), energy storage batteries, carbon capture and storage (CCS), and energy-saving household appliances. For example, while cobalt and graphite are required for batteries, wind and solar photovoltaic (PV) energy technologies rely on a variety of materials including neodymium, praseodymium, dysprosium, indium, gallium, and silicon metal. From this point of view, it is quite clear that raw materials are not only considered to be important due to economic or supply chain concerns but also because they have a substantial potential to affect the transition from a fossil-based energy system to another energy system powered by low-carbon energy technologies (Dominish et al., 2019; European Commission, 2020a; International Energy Agency, 2022a; Sovacool et al., 2020; United Nations Environment Programme International Resource Panel, 2020; Vidal et al., 2013).

Figure 1.1 shows the evolution of global domestic material consumption (DMC) and material footprint (MF) over the period between 1970 and 2019.[1] In line with the discussions above, the amount of material consumption measured in terms of both DMC and MF in 2019 is about 3 times higher than in 1970. Considering this increasing trend, it can also be projected that global demand for raw materials will grow considerably in the next few decades not only due to the production process and needs of the modern economy itself but also especially for ensuring the clean energy transition. For example, the transition from Internal Combustion Engine (ICE) cars to EVs is regarded as an essential component of the decarbonization of the road transport sector (Alataş, 2022; European Environment Agency, 2021; Hill et al., 2019).[2] However, the raw material (e.g., copper) used for EV production is about 4 times higher than a standard ICE. This also holds for the batteries, central to all EVs (Dominish et al., 2019; International Energy Agency, 2022a; Jones et al., 2020; Månberger & Stenqvist, 2018). Similar to EV production, PV systems and wind power plants require 11–40 times more copper and 6–14 times more iron, clearly indicating that the clean energy transition needs more materials (Hertwich et al., 2015).

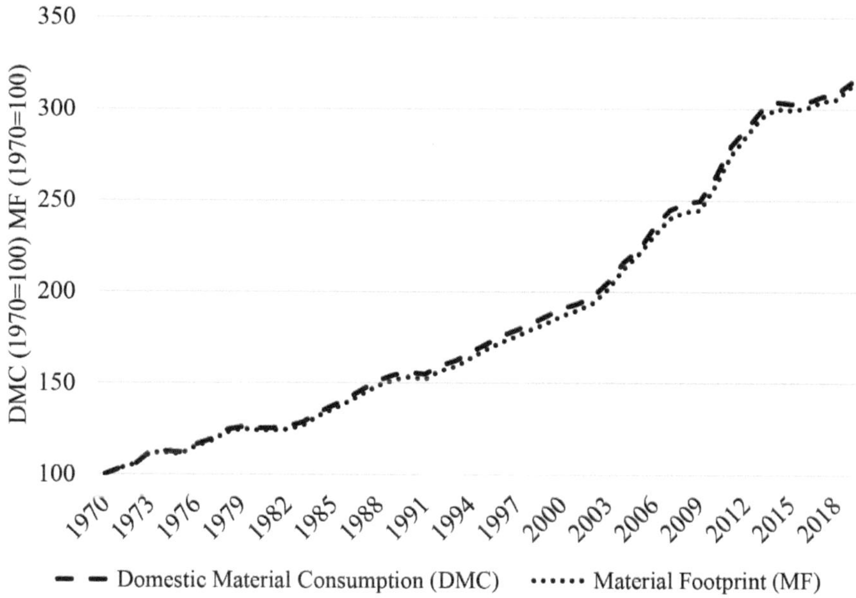

Figure 1.1. Domestic Material Consumption (DMC) and Material Footprint (MF) (1970–2019).

Source: United Nations Environment Programme International Resource Panel (2020).

The growing demand for materials is also evidenced by international organizations. For example, the Global Material Resources Outlook report of the Organization for Economic Cooperation and Development (Organisation for Economic Co-operation and Development, 2019) forecasts that global material use will increase from 79 billion tons in 2011 to 167 billion tons in 2060, implying about a 110% increase in global material use.[3] More importantly, non-metallic minerals and metals consist of more than 60% of this global use. Similarly, the World Bank suggests that demand for some raw materials that are intensively employed in the energy battery storage technologies, such as cobalt and lithium will rise by more than 1000% by 2050 (World Bank, 2017). Under the sustainable development scenario, International Energy Agency (2022a) projects that the share of clean energy technologies in total demand for lithium, cobalt, and nickel will increase by more than 60% by 2040, as the energy transition accelerates.

Despite playing a highly crucial role in securing the clean energy transition through the development of renewable energy or some other high technological advances—or in more general terms despite their strategic, economic, environmental, and technological importance—some concerns exist about the sustainable supply of raw materials. In this regard, it is argued that the raw materials supply may not be able to respond quickly to this increasing demand due to some reasons, which in the end might lead to some supply restrictions and slow the clean energy transition (European Commission, 2020a; Vidal et al., 2013). To be more specific, the Democratic Republic of Congo is estimated to have more than 50% of the known global cobalt reserves at the end of 2021, which is followed by Australia with about 20%. It means that more than 70% of global cobalt reserves are governed by two countries. Similarly, Chile hosts an estimated 10 million tons of lithium-about 45% of the world's known lithium reserves (British Petroleum, 2022). Therefore, as some materials are geographically concentrated in a small number of countries, this might pose a supply risk.

Needless to say, the supply risk of raw materials is not only related to the geographical concentration of reserves of those materials. There exist some other factors that increase the supply risk: environmental issues (energy-related emissions due to extraction or mining, pollution, waste) and social issues (weak labor protection, child labor), declining resource quality (lower-grade ores), production costs, and climate risks (International Energy Agency, 2022a). Therefore, the supply side of the raw materials might create some risks to be addressed, as the demand for clean energy transition requires a shift from abundant raw materials to those that are less abundant or at higher risks (Koyamparambath et al., 2022).

Due to all these issues discussed above, from the economic and strategic importance of raw materials to the higher risk of supply or price volatility, some of the raw materials needed for low-carbon technologies are considered critical. Therefore, they are given a special name in literature: critical raw materials (CRMs). These CRMs, as partially introduced previously, include cobalt, lithium, graphite, and rare earth elements, just to name a few (Jones et al., 2020; Pommeret et al., 2022; Watari et al., 2020).

CRMs are currently one of the most highlighted topics in academic research and environmental policy decision-making. As countries continue to demand more of such critical materials, this topic is expected to play a key role across the globe in the next decades not only in climate change mitigation or green transition issues but also in many areas from supply chain management to modern technological progress. In this chapter, we seek to understand the crucial role of CRMs by paying special attention to their role in the clean energy transition. For that purpose, we first briefly discuss the economic importance and supply risk of the CRMs from the green energy transition perspective and then point out some raw materials at supply risk and key clean technologies requiring these raw materials intensively.

The extraction of the CRMs from nature with building a mining project initially seems to be the most efficient policy for mitigating the supply risk, yet it might not be as it seems. This is because supply risk is not restricted to itself and might also bring along some other risks, such as environmental or social risks. More importantly, it might not be cost-efficient and not provide expected outcomes in the short term. Therefore, we secondly study circular economy strategies as a solution to supply risk as it significantly reduces the need for CRMs extraction from nature and dependence on unstable supply. In this regard, we draw our attention to the two CE strategies, i.e., recycling and reusing, focusing on their advantages and limitations.

It is worth noting that the CE strategies do not come without challenges and costs, and thus may not be alone sufficient to mitigate the supply risk. In this regard, the green energy transition and global investments in CRMs must be strongly supported by consistent and coherent economic policy instruments. Therefore, we third focus on the global clean energy investment gap and needs, possible financial and economic policies, and the role of stakeholders, i.e., governments, firms, and investors. The organization of the chapter is as follows. In the next section, we discuss the economic importance and supply risk of the CRMs for the clean energy transition. While Section 3 presents the CE strategies for mitigating the supply risk of CRMs, Section 4 discusses the global clean energy investment challenges and potential steps to address the ongoing supply gap in the CRM sector. Section 5 concludes.

CRITICAL RAW MATERIALS

CRMs are crucial in many respects. First, CRMs are linked to many industries, meaning that many different industries employ various raw materials as input in different amounts at various supply chain stages. As industrial production heavily relies on building resilient supply chains, any supply chain interruption caused by any factor, such as pandemics, natural disasters, trade regulations, war, or civil conflict might significantly lead to serious damage to the whole economy. This mainly occurs when there exists a mismatch between demand and supply. Second, CRMs underpin modern society by shaping technological progress. Many different technological products, applications, or services that we use intensively in our daily lives contain a wide variety of metals and minerals.[4] Third, not only the technological products that we use frequently in our daily lives but also modern clean technologies used in energy production such as wind turbines, EVs, or solar panels are closely related to CRMs. It means that the decarbonized future increases demand for new low-carbon technologies, and these technologies are more material-intensive, which essentially requires a substantial amount of various raw materials. Growing dependence on these raw materials is, therefore, quite obvious when it is assessed from a historical perspective. For example, the energy production industry mainly built on mills or steam engines in the 1700s and 1800s needed only a few materials, such as iron, carbon, copper, tungsten, or manganese. However, especially starting from the 2000s, the number and variety of materials (i.e., cobalt, lithium, or aluminum) employed in energy production increased significantly due to the different forms of energy production technologies (European Commission, 2018, 2020a; Jones et al., 2020; Watari et al., 2020; Zepf et al., 2014).

Considering the cruciality of the issue and growing concerns, countries, intergovernmental organizations, and academics have paid significant attention to determining raw materials criticality, designing some policies, and taking action on CRMs (Jones et al., 2020; Pommeret et al., 2022; Watari et al., 2020). For example, the EU has published a list of critical raw materials every three years since 2011. The first list contains only 14 materials. However, this number has gradually increased over the years, reaching 20 and 27 raw materials in 2014 and 2017, respectively. The last report published in 2020 identified 30 CRMs. Table 1.1 reports these CRMs (European Commission, 2020b). Similarly, the recent list of the United States Geological Survey (USGS) identifies 50 minerals critical for the US economy, which revises the first list published in 2018 by adding 15 more minerals (The United States Geological Survey, 2022a).[5] Undoubtedly, whether a raw material is critical or not is not only limited to these two lists published by the EU and US. There are many criticality assessments in the existing literature conducted at different levels, such as company, product, or technology (Schrijvers et al., 2020). As the purpose of these criticality assessments is diverse, they considerably differ from each other in many perspectives, such as method, dataset, identification, and outcome (Dewulf et al., 2016; Erdmann & Graedel, 2011; Graedel & Reck, 2016). In this chapter, without further elaboration on these technical issues,[6] we simply follow the EU criticality assessment. According to the methodology for the EU list of CRMs (European Commission, 2017), two parameters determine whether a raw material is critical or not: high economic importance and high supply risk. We discuss the meaning of these parameters for the clean energy transition in the following subsections.

Table 1.1. The EU List of Critical Raw Materials

Antinomy	*Borate*	*Germanium*	*Natural Graphite*	*Scandium*	*Tungsten*
Baryte	Cobalt	Hafnium	Natural Rubber	Silicon Metal	Strontium
Bauxite	Coking Coal	Indium	Niobium	Tantalum	HREEs
Beryllium	Fluorspar	Lithium	Phosphate Rock	Titanium	LREEs
Bismuth	Gallium	Magnesium	Phosphorus	Vanadium	PGMs

Note: According to the European Commission (2020b), HREEs, LREEs, and PGMs denote heavy rare earth elements, light rare earth elements, and platinum-group metals, respectively. While HREEs are dysprosium, erbium, europium, gadolinium, holmium, lutetium, terbium, thulium, ytterbium, and yttrium, LREEs are cerium, lanthanum, neodymium, praseodymium and samarium. PGMs are iridium, palladium, platinum, rhodium, and ruthenium.

ECONOMIC IMPORTANCE OF CRMS FOR THE CLEAN ENERGY TRANSITION

Being economically important mainly refers to the fact that those raw materials are essential for a wide range of products and technologies and are linked to all supply chain stages. As discussed previously, these products or technologies include not only facilities integrated into our daily lives, such as cell phones, computers, household appliances, or televisions but also low-carbon technologies, such as wind turbines, EVs, or PV systems. In this regard, the high economic importance of raw materials is also closely related to having irreplaceable environmental importance as they play a crucial role in the clean energy transition (Center for Sustainable Systems, 2021; European Commission, 2020b; The United States Department of Energy, 2011).

Table 1.2 simply summarizes some key clean technologies and their raw material requirements. As clearly seen, the number and type of materials largely vary across clean energy technologies, even within the same technology.[7] However, some materials become more prominent as they are needed for almost all clean technologies. These are copper, cobalt, nickel, graphite, lithium, aluminum, steel, and some REEs. It is equally important to note that even if a material is not listed as critical, it might be highly important for the given low-carbon technology. For example, copper is not classified as critical by the EU. Yet, both wind and solar PV technologies require copper. Therefore, Table 1.2 does not only include CRMs, but also raw and processed materials.

Low-carbon technologies and their material needs presented in Table 1.2 are also intercorrelated with each other. For example, hydrogen is the most abundant element in the universe, but its release from the compounds (i.e., water molecules) requires energy. Therefore, it is very important whether this energy comes from fossil fuels (grey or blue hydrogen) or renewable energy (green hydrogen). It means that more green hydrogen compatible with moving toward net zero emissions requires more renewable energy sources and their associated materials (International Renewable Energy Agency, 2022).

SUPPLY RISK OF CRMS FOR CLEAN ENERGY TRANSITION

As stated above, the economic importance of raw materials is mostly about supply chain management or technological applications. Supply risk is, on the other hand, mainly related to whether supply meets this growing demand. In other words, supply risk occurs in some cases, such as an increase in demand,

Table 1.2. Summary of Some Selected Low-Carbon Technologies and Materials Requirements

Technology	*Some Required Materials*		*Reference(s)*
Solar PV	Indium	Tellurium	(Dawkins et al., 2012)
	Gallium	Copper	(Elshkaki & Graedel, 2013)
	Silicon Metal	Aluminum	(European Commission, 2018)
	Selenium	Silver	(Grandell et al., 2016)
	Cadmium		(Giurco et al., 2019)
			(Carrara et al., 2020)
Wind Power	Neodymium	Carbon	(Dawkins et al., 2012)
	Dysprosium	Iron	(Elshkaki & Graedel, 2013)
	Praseodymium	Aluminum	(Grandell et al., 2016)
	Copper	Steel	(European Commission, 2018)
	Limestone		(Giurco et al., 2019)
			(Carrara et al., 2020)
EVs	Neodymium	Gallium	(European Commission, 2018)
	Dysprosium	Steel	(Giurco et al., 2019)
	Niobium	Graphite	(Carrara et al., 2020)
	Praseodymium	Cobalt	(Grandell et al., 2016)
Batteries	Neodymium	Copper	(European Commission, 2018)
	Dysprosium	Aluminum	(Giurco et al., 2019)
	Nickel	Steel	(Carrara et al., 2020)
	Manganese	Lithium	(Dawkins et al., 2012)
	Cobalt	Graphite	
LEDs	Gallium	Cerium	(Grandell et al., 2016)
Semiconductor	Indium	Europium	(Wilburn, 2012)
	Arsenic	Lanthanum	
	Yttrium		
Nuclear	Europium	Holmium	(Peterson et al., 2005)
	Samarium	Concrete	(The United States Geological Survey, 2022a)
	Hafnium	Steel	
	Copper	Nickel	(International Energy Agency, 2022a)
Hydro	Copper	Nickel	(International Energy Agency, 2022a)
Bioenergy Geothermal	Aluminum	Chromium	
	Zinc		
Hydrogen	Nickel	Aluminum	(International Energy Agency, 2022a)
Fuel Cells	PGMs		
Electricity Networks	Copper	Aluminum	(International Energy Agency, 2022a)

Note: PGMs denote platinum-group metals.

limitations that may arise in expanding global production capacity in the short and medium term, the structure of supply which is owned by monopolists or a few countries, or political risks (Moss et al., 2013: 559). Therefore, there are various indicators of supply risk, such as country concentration, country risk, depletion time, by-product dependency, commodity prices, mine/refinery capacity, risk of strategic use, abundance in the earth's crust, etc. (Achzet and Helbig, 2013).

Supply risk is not only restricted to the distribution of the CRMs reserves on earth and their extraction from nature. Heavily dependence on some specific countries for processing operations can also play a crucial role and might negatively affect supply. For instance, China is the dominant country in processing. The country's share of refining is around 90% for REEs, 60% for lithium, 65% for cobalt, 40% for copper, and 35% for nickel in 2019 (International Energy Agency, 2022a: 13). This means that China has started to monopolize significantly in terms of processing. America is an important example as it is dependent on imports for most of the nonfuel mineral products used in many technological goods from consumer goods to defense products. Unsurprisingly, its largest supplier is China. While the percentage of net import reliance of America is 100% for materials such as graphite, indium, vanadium, and yttrium, it is around 76% for cobalt, 48% for nickel, 45% for copper, 44% for aluminum, and 25% for lithium in 2021 (USGS Mineral Commodity Summaries, 2022b: 7).

Some environmental issues, such as energy-related emissions due to extraction or mining, pollution, and waste might also increase the supply risk (International Energy Agency, 2022a). It is because, as explained previously, various countries and companies are increasingly striving to use clean technologies requiring CRMs for the net zero target. Therefore, this trend increases the demand for critical materials, leading to environmental and supply risks. For example, according to the Organisation for Economic Co-operation and Development (2019) report, global material use is estimated to more than double in 2060 compared to 2011, and therefore greenhouse gas emissions from materials will rise to approximately 50 Gt CO_2 equivalent by 2060. According to this projection, 21% of the total emissions in 2060 will be caused by material extraction and processing. Besides, primary copper and nickel are the materials with the highest cradle-to-gate per kilogram environmental impact. It is equally important to note that not only material use through extraction and process but also existing material stocks contribute significantly to GHG emissions.

On the other hand, the mining of critical materials generates serious social risks. The best example of this issue is the Democratic Republic of the Congo, which has struggled to survive years of armed conflict. The majority of cobalt comes from this country, where women and children work for lower wages than men and lack basic safety equipment. Due to the lack of basic precautions, many cobalt miners have extremely high levels of toxic metals in their bodies (Sovacool et al., 2020). This situation, as with the other factors above, threatens the sustainability of supply and poses a risk.

Supply risk is accompanied by several different problems, i.e., price fluctuations. Historically, the most important example showing that supply risk will cause uncertainty and insecurity is the cobalt crisis that broke out in Zaire in 1979. In this cobalt crisis, the price of cobalt was $55,000/ton and dealer prices were reported to be $99,000/ton (Alonso et al., 2007). Hence, volatility in prices is generally considered one of the most important outcomes of supply risk. To observe a more recent example, according to International Energy Agency (2022a: 39), monthly standard deviations in price volatility of palladium and lithium between January 2010 and February 2021 are around 125% and 80%, respectively. In a similar vein, if we take the February 2015 value of the price index level of cobalt as a base year, this index increased to 300 in April 2018 and decreased below 100 in July 2018. Therefore, cobalt prices can also be considered quite volatile. Beyond price volatility, supply shortages can also lead to a variety of other challenges, such as production bottlenecks and the inability to deliver products on time.

STRATEGIES FOR MITIGATING SUPPLY RISK

As discussed in previous sections, one of the most important elements required for the net-zero target is technological transformation. This global energy transition will significantly trigger CRMs demand due to its important role in technological products.[8] This transformation brings along vital issues that need policy recommendations, such as supply risks, social risks, environmental risks, and technological risks. In this section, we focus only on supply risks and measures against these risks.

Circular economy (CE) appears as one of the most important strategies in policymaking to address supply risk, as it significantly reduces the need for CRMs extraction from nature and dependence on unstable supply (Gaustad et al., 2018; Babbitt et al., 2021; Karali and Shah, 2022). Materials can be resold as secondary raw materials to reduce the pressure on the primary input. When products containing CRMs come to the end of life, these products can be processed to recover materials, or their various parts can be repaired or reused in the same or different types of products. These solutions are known as CE strategies.

An important CE strategy that can be taken against supply risk is recycling. Recycling can reduce supply risk in two ways: a reduction in total imports (reduction effect) and redistribution of the import supply mix (redistribution effect) (Santillán-Saldivar et al., 2021). However, although it may reduce the supply risk, for many energy transition metals, such as lithium and REEs, recycling techniques are not well established. For example, copper is an essential element for electrical cables and devices and does not lose its physical and chemical properties when recycled. However, only 38% of copper was used from recycled sources in 2020 (Center for Sustainable Systems, 2021). In addition, recycling is not implemented as effectively as expected in some technologies, i.e., magnets. The stocks in global use for the four REEs (neodymium, praseodymium, terbium, and dysprosium) in the neodymium-iron-boron permanent magnet are as follows, respectively: 62.6 kilotons (kt), 15.7 kt, 3.1 kt, and 15.7 kt (Du and Graedel, 2011). Being able to recycle them effectively is crucial in relieving pressure on their primary supply, as their stocks were almost four times their annual extraction rate in 2007 (Yang et al., 2017).

With the advanced collection and recycling method, it may be possible to meet 37–91% of the demand for CRMs in low-carbon technologies from secondary materials in 2050 (Karali and Shah, 2022). According to the projections of the International Energy Agency (2022a:15), the primary supply requirement for copper, lithium, nickel, and cobalt could be reduced by around 10% with the recycling and reuse of electric vehicles and storage batteries. According to the report, this does not mean giving up the primary source entirely but will bring significant security and environmental benefits. It is because unlike developing and building a mining project taking between 8 to 15 years to fully operate, one of the main advantages of the CE strategies is that they are cost-efficient and can be immediately implemented in the short-term. However, it is equally important to note that as the recyclability of materials is different from each other, it might pose some limitations to the implementation of this strategy. REEs such as lanthanides, scandium, and yttrium, as well as tantalum, gallium, and indium have total end-of-life recycling rates of less than 1% (Buchert et al., 2012; Graedel et al., 2011).

Another limitation is that the products containing CRMs are not well-designed to upgrade, reuse, or disassemble, which makes it hard to access the critical materials inside them at the end of their life (Babbitt et al., 2021). However, the circular economy proposes to solve these limitations with the "slowing, closing, narrowing resource loops" approach introduced by Bocken et al. (2016). While the concept of slowing resource loops expresses the design of long-life goods and product life extension, the concept of closing resource loops refers to recycling. On the other hand, narrowing resource loops is used in the same sense as resource efficiency which is defined as reducing the resource used per product (Bocken et al., 2016: 309). These CE strategies can be extended. For example, strategies including dematerializing products, substituting digital alternatives for physical products, sharing or leasing products and services can be classified as the narrowing resource loops approach, while creating easily disassembled and upgraded products, repairing and remanufacturing can be grouped under the slowing resource loops approach (Babbitt et al., 2021).

Reusing is another important CE strategy that can be used to mitigate the supply risk of CRMs. In this regard, European countries' achievements can be a good example to understand the crucial role of reusing strategy. For example, in Europe in 1996, 51% of used tires were either reused as second-hand tires, reconditioned through retreading, recycled, or sent for energy recovery. Over the years, the organization of natural rubber obtained from used tires performed well in Europe and this share increased to 92% in 2015. However, it is still weak in the reuse and second use of tires (European Commission, 2018: 50).

The CE strategies do not come without challenges and costs. In this regard, for example, an increase in costs from the implementation of CE strategies might create some barriers to the greater uptake of CE strategies. In a similar vein, some CRMs are employed more in the production process as their prices are lower. These issues can result in two ways: i) incentivizing recycling, recovery, design for recycling, etc., and ii) price increase with rapid climbing of demand and associated stress in supply chains. Innovation and R&D are one of the most important solutions to overcome commercial barriers. In this regard, designing product-specific collection schemes, designing a special e-waste product category, executing legislative action that incentivizes collection and recycling, tracking collection and recycling activities, and designing penalty and reward mechanisms are some other suggested potential solutions (Karali and Shah, 2022).

The recycling share of CRMs or REEs on a global scale is quite low and it is still at an early stage. This low recycling rate level is clearly evidenced by REEs with less than 1% (Center for Sustainable Systems, 2021). Although it is currently recycled at a very low rate, there is a strong possibility that recycling will become more dominant soon. This will be possible thanks to the growing waste of clean energy technologies. As the deployment of many clean energy technologies will be accelerated in the coming years and as they contain a significant amount of critical raw materials, applying circular economy strategies by collecting these wastes will become more feasible (International Energy Agency, 2022a: 15). It is possible to exemplify this with technologies such as wind power, solar PV and batteries. If these technologies reach the end-of-life stage, the critical raw materials they contain have a high potential for recycling. However, considering the issues discussed above, it is quite clear that a circular economy alone is not sufficient to mitigate the supply risk, as the recyclability, reusability, etc. of materials are different.

There are also different solutions not included in the scope of CE strategies, but that are thought to be a solution to a larger part of the supply risk. Research and development (R&D), developing diplomatic strategies, and diversification are the most common examples of these types of solutions. The R&D strategy mainly refers to the reduction of the demand for CRMs by paying attention to technical issues such as efficiency, recycling, and the development of substitutes. CRMs are also the subject of diplomatic strategies as they are closely correlated with policy, economics, and defense. Methods such as information sharing, agreements, collaborations, and trade law reforms are examples of these diplomatic strategies. It is possible to achieve the last strategy, called diversification, by expanding supply networks (new upstream mineral producers) and broadening the number of players in the mid-stream processing stage (Wilson, 2020). Many of these strategies also could have positive secondary benefits for firms, and factors such as energy consumption, waste, pollution, and costs could be reduced in this way. The decrease in these costs may also be one of the impulsive reasons for the incentives and implementation of strategies for firms (Gaustad et al., 2018). How all this can be achieved will be discussed in the next section from the perspective of investment and policy discussion.

Both circular economy and non-circular economy measures need to be applied on a case-by-case basis. To look at the country-specific example, the dependence of the European Union on China for tellurium, indium, and gallium in the solar industry and for neodymium and dysprosium in the wind industry represents a case where the supply risk cannot be mitigated only with circular economy measures. In this regard, there are some other measures to be taken by both the European Union and the producers. Among the measures that the European Union can take are bilateral negotiations, the promotion of research within the EU, and also with international partners by using the dialogue on international platforms such as the G20 and OECD.

KEY INVESTMENT CHALLENGES, POTENTIAL STRATEGIES, AND SOME POLICY OPTIONS

Clean energy transition is one of the greatest challenges of the current times and it is important to ascertain to what extent current and projected levels of investments are mobilized/needed to meet the requirements of such a great challenge. There are a number of international organizations that annually calculate and report such information. Recent reports suggest that global clean energy investments continued to increase, despite the COVID-19 crisis and Russia's invasion of Ukraine. For example, according to the Bloomberg New Energy Finance (BNEF) report, energy transition investment and climate-tech corporate finance totaled 920 billion USD in 2021, indicating an over 25% increase compared to the previous year. With a 60% increase in investment compared to the previous year, China's clean energy investment reached 266 billion USD, making it the leader in clean investments, followed by the USA, the EU countries, and Japan (BNEF, 2022). IEA World Energy

Investment report also highlights that a 1.4 trillion USD investment is expected in clean energy technologies in 2022 (International Energy Agency, 2022b).[9] However, the current increase in global clean energy investment is far behind what is needed for 2050 net-zero targets. To get on track for global net zero targets, according to the BNEF estimates, energy transition investments must triple by 2025, and then double by 2030 (BNEF, 2022). In a similar vein, IEA projects that the annual investments in real terms must be increased to 5 trillion USD in 2035, then slightly falling to 4.5 trillion USD in 2050 (International Energy Agency, 2022b).

As we have mentioned earlier, clean-energy technologies are more material-intensive to build than their fossil-fuel counterparts, and therefore a significant amount of CRMs are needed for decarbonization. The investment picture, on the other hand, for the mining industry is more gloomy. Investment in the mining industry already had been falling since 2012. Data indicates that investments in basic metal and mining companies have fallen from 120 billion USD to below 50 billion USD in 2016 and stabilized around these levels since then (Wood Mackenzie, 2022). Besides, the prices of basic metals in general and CRMs in particular increased significantly since the beginning of 2021, mainly due to rising demand, disrupted supply chains, and concerns around tightening supply. Such rapid increases in recent CRMs prices have both negative and positive outcomes for the clean energy transition. As for the negative consequences, price surges in critical materials threaten to reverse the long-term trend in the falling costs of clean energy technologies as many clean energy technologies are far more material intensive than conventional energy sources. IEA, for instance, estimates that while the share of cathode materials in the total cost of the lithium-ion battery pack was less than 5% around five years ago, their share has increased to 20% in the first quarter of 2022 (International Energy Agency, 2022b; Kim, 2022). On the other hand, due to price surge, returns and margins have been higher in commodities markets in general and CRMs in particular up until the second quarter of 2022, current market fundamentals are in favor of encouraging investments in CRMs. IEA World investment report in 2022, for the first time, estimated that total capital expenditures (CAPEX) for selected CRMs reached above 33 billion USD in 2021, indicating a 30% increase compared to the previous years (International Energy Agency, 2022b). Despite supply chain issues and centralized production of CRMs in certain regions, current supplies of CRMs seem sufficient to meet the current demand for clean energy sources, yet the numbers are well short of what is required for net-zero emission targets. Based on recent estimates, around 400 billion USD investment is needed in the CRMs industry until 2030 globally to support the deployment of renewable energy, electric vehicles, and battery needs (Kettle, 2022). It is particularly worrying as mining investments have been flat for a long period and stayed well below 2012 figures, as stated above. Moreover, depending on the type of critical minerals, developing a mining project from the scratch takes between 10 to 15 years to fully operate (Umbach, 2021; International Energy Agency, 2022a). It is also worth noting that a steadily increasing cumulative scale of CRMs is needed to meet the challenge of the low-carbon transition. Therefore, it is urgent to mobilize and accelerate global investments in CRMs with consistent and strongly incentivized policies and measures.

While the demand will increase quite significantly in the coming years, the supply will be well short of what is needed for the same period (Kettle, 2022). The supply gap that is projected for after 2030 is enormous. However, it also presents a great opportunity for the investors as increasing demand will ensure that the market will be huge and profitable for such investment in CRMs. Therefore, contrary to the fossil industry, the CRM industry has multi-decade sights as to the direction of investment pathways as great energy transition secures and provides great opportunities for them.

As for green energy transition, CRMs are essential and the projected supply gap indicates that countries should immediately step up their efforts to combat the climate crisis by ramping up investment. In this regard, international collaboration is an extremely important topic, in terms of enhancing trade and investment relationships with a variety of partners. Collaboration should be done both among and between developed and developing countries in a number of priority areas. When a developing country finds a significant amount of CRMs reserves, it might not be possible to extract and process these mining reserves simply because of technical difficulties, lack of investment, or know-how. For diversification purposes, Western developed countries should financially and technologically support such projects and secure the supply chain of CRMs. Such successful collaboration is already seen in the case of the Lynas project. Lynas is a rare earth project in Western Australia, producing neodymium and praseodymium from its Mt. Weld mine and then shipping them to Malaysia for processing. The project is almost fully financed by a government-backed Japanese consortium (Wood Mackenzie, 2022). Similarly, Turkey has recently announced a significant amount of REE in their provinces and such an effective partnership could be developed with the Western countries to secure and diversify the supply chain of CRMs. Besides, since the investment requirement is so substantial and urgent, and further grow more in the coming years, it is not convenient to rely only on public funding as it will not be enough to meet such requirements. Therefore, governments must encourage funding to be supplied for such investments from multilateral organizations, financial institutions, and the corporate sector.

Investment and operations should aim at every stage of the supply chain. The refining and processing of CRMs are concentrated mainly in some specific countries, such as China. Thus, focusing only on mining exploration will not add to higher supply security. As building mining, refining, and processing plants require a large amount of upfront capital, e.g., many times more than a billion USD, such investments can be done either by governments or big companies. Market forces alone won't solve the problem and, therefore, the responsibility is on governments to take the lead. To convince and encourage the private sector to invest in long-term mining, the governments must ensure fair, long-term prices and demand certainty.

The role of governments in CRMs investment and operations may become more important because of China's current dominance and the way the CRMs and RREs are developed and processed in China. China has seen the strategic importance of CRMs well before the West and heavily invested in these minerals and REEs at the state level without paying much attention to the profit-cost dimension and environmental consequences of such activities. Moreover, with strong industrial policy support and research, China gained a dominant position at all levels of CRMs production, including from mining to refining and processing, giving them a competitive advantage in the medium term (Leiden-Delft-Erasmus Universities, 2022). Even more than mining and extraction, due to strategic actions taken earlier, China is currently more dominant in processing operations. For instance, China's share of refining is around 35% for nickel, 60% for lithium, 65% for cobalt, and almost 90% for REEs, as indicated earlier (International Energy Agency, 2022a). It is fair to argue that China has gained such overwhelming CRMs market control by granting only state-controlled enterprises to operate in this field and providing a significant amount of investment. It is, therefore, imperative for the Western governments to take the lead and operate in the CRMs industry at the state level as the current environment, social and governance (ESG) rules, and market conditions make it more challenging to compete with China. With stringent environmental regulations and higher labor costs, the private sector in the West will seriously feel this enormous challenge and may be discouraged from investing. In this regard, public-private partnership (PPP) in CRMs investments could be an effective solution to overcome such barriers. A certain level of state support and intervention is needed, and the main advantage of a PPP contract lies in the risk-sharing of the governments with the private company.

For the private sector to make a long-term investment, they have to be ensured that their products will be demanded and bought at a fair price. As new kinds of mining and processing of CRMs are expected to be produced in line with ESG rules, they deserve a secure market demand and a "green premium." To achieve this, governments could act consistently to create such demand through green public procurements or regulated standards for such environmentally friendly productions. The US government, for instance, recently introduced such guarantees for the production of CRMs produced locally.[10] Similarly, private corporations with net-zero targets can also stimulate such demand for cleaner CRM producers as we have seen such actions under the RE100 initiative, where large corporations are sourcing renewable energy without a regulatory obligation (Delbeke, et al., 2022).

As for the big private companies, it is essential to re-consider and restructure the way they make their business plans and strategies with newly growing technological developments. At least until the COVID-19 pandemic times, mass production of auto-industry for combusted fuel engine vehicles did not have to worry about getting their supplies from different countries as such parts and supplies were readily available. Yet it is not the case for EV producers as securing critical parts and their batteries are far more difficult and scattered only in limited areas. If they do not secure such a supply as of today, the risk will be much higher in the future. It might be a wise strategy for current EV companies to invest not only in downstream but also in upstream projects. In this regard, acquiring part of their supply chain (vertical integration) as a means to mitigate supply chain risk could be a better solution. It is important to remember that Henry Ford in earlier times focused on vertical integration for automotive production and invested in rubber plantations in Brazil and iron and steel in Detroit, which gave his company the scale needed to improve performance and lower costs. Current EV companies could learn from this experience and may look to move upstream and directly acquire new mining projects, which basically means buyers of critical materials actually buy and own the mines that supply them. We already see such moves. For example, Tesla, General Motors, and BMW invest in lithium, cobalt, and rare earth minings in some other parts of the world to safeguard their production pipelines (International Energy Agency, 2022b; Naor et al., 2021).

CRMs mean all about innovation. When the world works immensely on new technologies to address the needs of the green energy transition, they recognize the CRMs' crucial role. Technological innovation is an important way to diminish, substitute, or even radically change the demand for CRMs. Therefore, an increased focus on R&D in mining, refining and processing, recycling, and recovery technologies will be required to help mitigate CRMs supply risks. China has invested significantly in R&D at an earlier stage and that is how they worked to scale significant production of CRMs. Ramez Naam states that as R&D is the most effective policy tool in the early stages of technological innovations, an increasing level of public funding should be allocated for R&D (Naam, 2022). As for the rest of the world, CRMs processing is a relatively new business, and they are lagging behind China in terms of know-how and efficient use of these technologies. Therefore, the West has to develop a roadmap and invest significantly in R&D to develop resilient and secure critical mineral and raw material supply chains independent of China. Governments can also stimulate more private involvement in R&D by developing consistent policy tools and enabling environments to increase and mobilize private investments.

Apart from these, a diverse range of economic and financial policies and tools should be considered to enable the large-scale production of CRMs and take advantage of learning curves as fast as possible. These policies and measures should be targeted for both short-term and long-term purposes to provide the right signals for the private sector. In this way, private companies can make investments safely and ensure that their projects will be profitable. At the earlier stage of the project, direct grants could be considered a policy tool. During commercializing stages or when the investments are maturing, economic tools, such as low-interest rate financing, might reduce risks for investors, buy time to scale up technologies and programs, and achieve longer-term financial sustainability. Governments can also deploy tax incentives, targeted subsidies to scale new technologies, favorable lending, and sustainable finance like green bonds. As CRMs business requires long-term planning from mining to processing to product, it is essential to develop economic and financial instruments to meet the requirements of these needs. Fortunately, apart from governmental support, applications of climate finance tools and institutions developed to fund and implement these finances are well-developed and might serve the needs of private businesses to invest in CRMs projects.

CONCLUSION

Since adopting the United Nations Paris Climate Agreement limiting the temperature rise to 2°C, preferably 1.5°C relative to pre-industrial levels, a growing number of countries have committed to net zero emissions targets. Transitioning to net-zero emissions undoubtedly requires implementing many different climate policies that lead to rapid and deep reductions in emissions. In this regard, clean energy technologies are considered to play a critical role in achieving these net-zero targets. However, building clean-energy technologies is more material-intensive than their fossil-fuel counterparts, and therefore a significant number of CRMs are needed to make progress in decarbonization. This chapter investigates the crucial role of CRMs from the clean energy transition perspective. For that purpose, we first discuss the economic importance and supply risk of the CRMs. Then, we evaluate the key policy instruments to be implemented to mitigate the supply risk, paying special attention to the circular economy and clean energy investment strategies.

CRMs are considered to be crucial not only because they are used as input in producing a large number of products but also because they have a substantial potential to affect the transition from a fossil-based energy system to another energy system powered by low-carbon energy technologies. Despite their important role in securing the clean energy transition, as the supply of these raw materials is neither sustainable nor fast enough to meet this growing demand due to various reasons, this might lead to some supply risks and hinder the green energy transition. Factors such as the concentration of supply in certain countries, serious ESG concerns, the mine/refinery capacity of countries' CRMs, strategic uses of CRMs, depletion time, by-product dependency, and commodity prices are identified as the main reasons for the supply risk.

Circular economy strategies such as recycling, reusing, and remanufacturing appear as important strategies that can be applied in the short term to mitigate some of these supply risks. Although circular economy strategies are necessary, cost-efficient, and might have a much bigger role in the future, they are not sufficient on their own for various reasons. For example, not all CRMs have the same recyclability rate, which prevents this strategy from being applicable to all CRMs. Alternatively, we encounter other important strategies, such as diversification, focusing on R&D of efficiency, recycling, and the development of substitutes, developing diplomatic strategies, and international collaboration. The combination of both circular economy and these other strategies will help to ease the supply risks and at the same time reduce pressure on primary supply.

As the green energy transition is moving forward with ambitious net-zero targets, investment is one of the greatest challenges. The investment is particularly more of an issue for the CRM industry as the projected supply gap is widening and building a mining and processing plant usually takes more than a decade to fully operate. A significant amount of money is needed to cover upfront costs and this cannot be done without government guidance and support once considering China's current dominance in many CRMs activities. All stakeholders, including governments, companies, investors, and other parties, are wakening up to the challenge and urgency of the need to ramp up investment all around. To address such big challenges, an efficient and timely energy transition must be supported by consistent and coherent policy incentives for the development of CRMs. Increased investment efforts need to be supported by a comprehensive plan that considers recycling, supply chain resilience, technology innovation, and sustainability standards.

ACKNOWLEDGMENTS

This research is developed within the context of the research project entitled "Analysing Material Demand and Material Efficiency from Sustainability Perspective: A Comparative Cross-Country Analysis and Assessments for

Turkey" (Project Number: 221K081). The project is funded by TUBITAK (The Scientific and Technological Research Council of Turkey) "1001—The Scientific and Technological Research Projects Funding Program."

NOTES

1. While DMC mainly measures the total material used in the economy, MF is the total amount of raw materials embedded in the final consumption (Karakaya et al., 2021; Wiedmann et al., 2015).

2. Road transport depends heavily on fossil fuels. This sector, therefore, is one of the most important sources of CO2 emissions (Alataş, 2022; European Environment Agency, 2021; Hill et al., 2019).

3. This report does not specifically focus on the material requirement of clean energy technologies. The main attention is rather paid to the economic drivers and environmental consequences of material use, such as population, output, economic structure, and materials-related emissions (OECD, 2019).

4. For example, 50 different metals in different quantities are needed to produce a standard smartphone (European Commission, 2018).

5. The CRMs identified by the United States Geological Survey (2022a) for the US economy are as follows: aluminum, antimony, arsenic, barite, beryllium, bismuth, cerium, cesium, chromium, cobalt, dysprosium, erbium, europium, fluorspar, gadolinium, gallium, germanium, graphite, hafnium, holmium, indium, iridium, lanthanum, lithium, lutetium, magnesium, manganese, neodymium, nickel, niobium, palladium, platinum, praseodymium, rhodium, rubidium, ruthenium, samarium, scandium, tantalum, tellurium, terbium, thulium, tin, titanium, tungsten, vanadium, ytterbium, yttrium, zinc, and zirconium. Although the EU and US lists of CRMs are very similar to each other, there are some differences. For example, while borate is listed as a critical element by the EU (Turkey has 98% of the EU's borate supply), it is not considered critical by the US. Similarly, nickel is a critical material for the US economy, but it is not for the EU. Moreover, the United States Geological Survey (2022a) splits the REEs and PMGs into individual entries, rather than including them as a group.

6. The methods, indicators, or identification processes used in the criticality of the raw materials are out of scope of this study.

7. For example, energy storage systems can be grouped under some categories, such as mechanical, electrochemical, electrical, and so on. Each category's material needs widely vary from each other (Küfeoğlu, 2020). This also holds for solar PV technologies (Grandell et al., 2016).

8. For detailed information and statistics, please see Hund et al. (2020).

9. Please note that BNEF and IEA current investment figures do not match for the same years, because BNEF does not include energy efficiency and grid investment figures into their calculations.

10. See https://www.whitehouse.gov/briefing-room/statements-releases/2022/02/22/fact-sheet-securing-a-made-in-america-supply-chain-for-critical-minerals/ (last accessed 09/1/2023).

BIBLIOGRAPHY

Achzet, B., & Helbig, C. (2013). How to Evaluate Raw Material Supply Risks—An Overview. *Resources Policy*, 38(4), 435–447.

Alataş, S. (2022). Do Environmental Technologies Help to Reduce Transport Sector CO_2 Emissions? Evidence from the EU15 Countries. *Research in Transportation Economics*, *91*, 101047. https://doi.org/10.1016/j.retrec.2021.101047

Ali, S. H., Giurco, D., Arndt, N., Nickless, E., Brown, G., Demetriades, A., Durrheim, R., Enriquez, M. A., Kinnaird, J., Littleboy, A., Meinert, L. D., Oberhänsli, R., Salem, J., Schodde, R., Schneider, G., Vidal, O., & Yakovleva, N. (2017). Mineral Supply for Sustainable Development Requires Resource Governance. *Nature*, *543*(7645), 367–372. https://doi.org/10.1038/nature21359

Alonso, E., Gregory, J., Field, F., & Kirchain, R. (2007). Material Availability and the Supply Chain: Risks, Effects, and Responses. *Environmental Science & Technology*, 41(19).

Babbitt, C. W., Althaf, S., Rios, F. C., Bilec, M. M., & Graedel, T. E. (2021). The Role of Design in Circular Economy Solutions for Critical Materials. *One Earth*, 4(3), 353–362.

BNEF. (2022). *Energy Transition Investment Trends 2022: Tracking Global Investment in the Low-carbon Energy Transition.* https://about.bnef.com/energy-transition-investment/

Bocken, N. M., De Pauw, I., Bakker, C., & Van Der Grinten, B. (2016). Product Design and Business Model Strategies for a Circular Economy. *Journal of Industrial and Production Engineering*, 33(5), 308–320.

British Petroleum. (2022). *Statistical Review of World Energy/All Data, 1965–2021.* https://www.bp.com/en/global/corporate/energy-economics/statistical-review-of-world-energy.html

Buchert, M., Manhart, A., Bleher, D., & Pingel, D. (2012). Recycling Critical Raw Materials from Waste Electronic Equipment. *Freiburg: Öko-Institut eV*, 49(0), 30–40.

Carrara, S., Alves Dias, P., Plazzotta, B., & Pavel, C. (2020). Raw Materials Demand for Wind and Solar PV Technologies in the Transition towards a Decarbonised Energy System. In *EUR 30095 EN*. https://doi.org/10.2760/160859

Center for Sustainable Systems. (2021). *Critical Materials Factsheet*. https://doi.org/10.1016/c2018-0-04537-5

Dawkins, E., Roelich, K., Falk, R., & Chadwick, M. (2012). Metals in a Low-Carbon Economy: Resource Scarcity, Climate Change and Business in a Finite World. *Stockholm Environment Institute*. https://www.jstor.org/stable/resrep00423

Delbeke, J., Egenhofer, C., & Lamas, R. (2022). Options to Finance the Premium Cost of Climate-neutral Products in the EU: The Potential of the ETS and Demand Creation. *STG Policy Briefs*, 2022/16. http://hdl.handle.net/1814/74641

Dewulf, J., Blengini, G. A., Pennington, D., Nuss, P., & Nassar, N. T. (2016). Criticality on the International Scene: Quo Vadis? *Resources Policy*, *50*, 169–176. https://doi.org/10.1016/j.resourpol.2016.09.008

Dominish, E., Teske, S., Florin, N., & Teske, S. (2019). Responsible Minerals Sourcing for Renewable Energy. *Report Prepared for Earthworks by the Institute for Sustainable Futures, University of Technology Sydney*. https://www.earthworks.org/publications/responsible-minerals-sourcing-for-renewable-energy/

Du, X., & Graedel, T. E. (2011). Global Rare Earth in-use Stocks in NdFeB Permanent Magnets. *Journal of Industrial Ecology*, 15(6), 836–843.

Elshkaki, A., & Graedel, T. E. (2013). Dynamic Analysis of the Global Metals Flows and Stocks in Electricity Generation Technologies. *Journal of Cleaner Production, 59*, 260–273. https://doi.org/10.1016/j.jclepro.2013.07.003

Erdmann, L., & Graedel, T. E. (2011). Criticality of Non-Fuel Minerals: A Review of Major Approaches and Analyses. *Environmental Science and Technology*, *45*(18), 7620–7630. https://doi.org/10.1021/es200563g

European Commission. (2017). Methodology for Establishing the EU List of Critical Raw Materials. In *Publications Office of the European Union*. https://doi.org/10.2873/769526

European Commission. (2018). Report on Critical Raw Materials and the Circular Economy. In *Publications Office of the European Union*. https://doi.org/10.2873/331561

European Commission. (2020a). *Communication from the Commission to the European Parliament, the Council, the European Economic and Social Committee and the Committee of the Regions: Critical Raw Materials Resilience: Charting a Path towards Greater Security and Sustainability*. https://eur-lex.europa.eu/legal-content/EN/TXT/?uri=CELEX%3A52020DC0474

European Commission. (2020b). Study on the EU's List of Critical Raw Materials-Final Report. In *European Commission*. https://doi.org/10.2873/11619

European Environment Agency. (2021). *Greenhouse Gas Emissions from Transport in Europe*. https://www.eea.europa.eu/ims/greenhouse-gas-emissions-from-transport

Gaustad, G., Krystofik, M., Bustamante, M., & Badami, K. (2018). Circular Economy Strategies for Mitigating Critical Material Supply Issues. *Resources, Conservation and Recycling*, 135, 24–33.

Girtan, M., Wittenberg, A., Grilli, M. L., Oliveira, D. P. S. De, Giosu, C., & Ruello, M. L. (2021). The Critical Raw Materials Issue between Scarcity, Supply Risk, and Unique Properties. *Materials, 14*, 1826. https://doi.org/https://doi.org/10.3390/ma14081826

Giurco, D., Dominish, E., Florin, N., Watari, T., & McLellan, B. (2019). Requirements for Minerals and Metals for 100% Renewable Scenarios. In S. Teske (Ed.), *Achieving the Paris Climate Agreement Goals: Global and Regional 100% Renewable Energy Scenarios with Non-Energy GHG Global and Regional 100% Renewable Pathways for +1.5°C and +2°C*. https://doi.org/10.1007/978-3-030-05843-2

Graedel, T. E., Allwood, J., Birat, J.-P., Reck, B. K., Sibley, S. F., Sonnemann, G., Buchert, M. & Hagelüken C. (2011). Recycling Rates of Metals: A Status Report. United Nations Environment Programme.

Graedel, T. E., & Reck, B. K. (2016). Six Years of Criticality Assessments: What Have We Learned So Far? *Journal of Industrial Ecology*, *20*(4), 692–699. https://doi.org/10.1111/jiec.12305

Grandell, L., Lehtilä, A., Kivinen, M., Koljonen, T., Kihlman, S., & Lauri, L. S. (2016). Role of Critical Metals in the Future Markets of Clean Energy Technologies. *Renewable Energy*, *95*, 53–62. https://doi.org/10.1016/j.renene.2016.03.102

Hertwich, E. G., Gibon, T., Bouman, E. A., Arvesen, A., Suh, S., Heath, G. A., Bergesen, J. D., Ramirez, A., Vega, M. I., & Shi, L. (2015). Integrated Life-cycle Assessment of Electricity-supply Scenarios Confirms Global Environmental Benefit of Low-carbon Technologies. *PNAS*, *112*(20), 6277–6282. https://doi.org/10.1073/pnas.1312753111

Hill, G., Heidrich, O., Creutzig, F., & Blythe, P. (2019). The Role of Electric Vehicles in Near-Term Mitigation Pathways and Achieving the UK's Carbon Budget. *Applied Energy*, *251*, 113111. https://doi.org/10.1016/j.apenergy.2019.04.107

Hund, K., La Porta, D., Fabregas, T.P., Laing, T., & Drexhage, J. (2020). Minerals for Climate Action: The Mineral Intensity of the Clean Energy Transition, World Bank Group, 2020. https://pubdocs.worldbank.org/en/961711588875536384/Minerals-for-Climate-Action-The-Mineral-Intensity-of-the-Clean-Energy-Transition.pdf.

International Energy Agency. (2022a). *The Role of Critical Minerals in Clean Energy Transitions*. https://doi.org/10.1787/f262b91c-en

International Energy Agency. (2022b). *World Energy Investment 2022*. IEA, Paris https://www.iea.org/reports/world-energy-investment-2022

International Renewable Energy Agency. (2022). *Geopolitics of the Energy Transformation: the Hydrogen Factor*. www.irena.org/publications

Jones, B., Elliott, R. J. R., & Nguyen-Tien, V. (2020). The EV Revolution: The Road ahead for Critical Raw Materials Demand. *Applied Energy*, *280*, 115072. https://doi.org/10.1016/j.apenergy.2020.115072

Karakaya, E., Sarı, E., & Alataş, S. (2021). What Drives Material Use in the EU? Evidence from Club Convergence and Decomposition Analysis on Domestic Material Consumption and Material Footprint. *Resources Policy*, *70*. https://doi.org/10.1016/j.resourpol.2020.101904

Karali, N., & Shah, N. (2022). Bolstering Supplies of Critical Raw Materials for Low-carbon Technologies through Circular Economy Strategies. *Energy Research & Social Science*, 88, 102534.

Kastanaki, E., & Giannis, A. (2022). Forecasting Quantities of Critical Raw Materials in Obsolete Feature and Smart Phones in Greece: A Path to Circular Economy. *Journal of Environmental Management*, *307*, 114566. https://doi.org/10.1016/j.jenvman.2022.114566

Kettle, J. (2022). Does the Energy Transition Start and Finish with Metals? *Wood Mackenzie Opinion*, https://www.woodmac.com/news/opinion/have-miners-missed-the-boat-to-invest-and-get-ahead-of-the-energy-transition/

Kim, T.Y. (2022). Critical Minerals Threaten a Decades-long Trend of Cost Declines for Clean Energy Technologies. *IEA Commentary*, https://www.iea.org/commentaries/critical-minerals-threaten-a-decades-long-trend-of-cost-declines-for-clean-energy-technologies

Koyamparambath, A., Santillán-Saldivar, J., McLellan, B., & Sonnemann, G. (2022). Supply Risk Evolution of Raw Materials for Batteries and Fossil Fuels for Selected OECD Countries (2000–2018). *Resources Policy, 75*. https://doi.org/10.1016/j.resourpol.2021.102465

Küfeoğlu, S. (2020). Emerging Technologies Value Creation for Sustainable Development. In *Encyclopedia of Earth Sciences Series*. Springer. https://doi.org/10.1007/978-0-387-36699-9_43

Leiden-Delft-Erasmus Universities. (2022). White Paper: Critical Materials, Green Energy and Geopolitics: A Complex Mix, Centre for Sustainability Leiden-Delft-Erasmus Universities. https://www.centre-for sustainability.nl/uploads/cfs/attachments/Critical%20Materials_ LDE%20White%20Paper_DEF20220627_0.pdf

Månberger, A., & Stenqvist, B. (2018). Global Metal Flows in the Renewable Energy Transition: Exploring the Effects of Substitutes, Technological Mix and Development. *Energy Policy, 119*, 226–241. https://doi.org/10.1016/j.enpol.2018.04.056

Moss, R. L., Tzimas, E., Kara, H., Willis, P., & Kooroshy, J. (2013). The Potential Risks from Metals Bottlenecks to the Deployment of Strategic Energy Technologies. *Energy Policy*, 55, 556–564.

Naam, Ramez (2022). Interview: Ramez Naam, Futurist, Author, and Investor. Noahpinion, April 23, https://noahpinion.substack.com/p/interview-ramez-naam-futurist-author

Naor, M., Coman, A., & Wiznizer, A. (2021). Vertically Integrated Supply Chain of Batteries, Electric Vehicles, and Charging Infrastructure: A Review of Three Milestone Projects from Theory of Constraints Perspective. *Sustainability*, 13(7), 3632.

Organisation for Economic Co-operation and Development. (2019). *Global Material Resources Outlook to 2060: Economic Drivers and Environmental Consequences*. https://doi.org/10.1787/9789264307452-en

Peterson, P. F., Zhao, H., & Petroski, R. (2005). *Metal And Concrete Inputs for Several Nuclear Power Plants* (Report UCBTH-05–001). https://fhr.nuc.berkeley.edu/wp-content/uploads/2014/10/05-001-A_Material_input.pdf

Pommeret, A., Ricci, F., & Schubert, K. (2022). Critical Raw Materials for the Energy Transition. *European Economic Review, 141*, 103991. https://doi.org/10.1016/j.euroecorev.2021.103991

Rabe, W., Kostka, G., & Stegen, K. S. (2017). China's Supply of Critical Raw Materials: Risks for Europe's Solar and Wind Industries?. *Energy Policy*, 101, 692–699.

Santillán-Saldivar, J., Cimprich, A., Shaikh, N., Laratte, B., Young, S. B., & Sonnemann, G. (2021). How Recycling Mitigates Supply Risks of Critical Raw Materials: Extension of the Geopolitical Supply Risk Methodology Applied to Information and Communication Technologies in the European Union. *Resources, Conservation and Recycling*, 164, 105108.

Schandl, H., Hatfield-Dodds, S., Wiedmann, T., Geschke, A., Cai, Y., West, J., Newth, D., Baynes, T., Lenzen, M., & Owen, A. (2016). Decoupling Global Environmental

Pressure and Economic Growth: Scenarios for Energy Use, Materials Use and Carbon Emissions. *Journal of Cleaner Production, 132*, 45–56. https://doi.org/10.1016/j.jclepro.2015.06.100

Schrijvers, D., Hool, A., Blengini, G. A., Chen, W. Q., Dewulf, J., Eggert, R., van Ellen, L., Gauss, R., Goddin, J., Habib, K., Hagelüken, C., Hirohata, A., Hofmann-Amtenbrink, M., Kosmol, J., Le Gleuher, M., Grohol, M., Ku, A., Lee, M. H., Liu, G., . . . Wäger, P. A. (2020). A Review of Methods and Data to Determine Raw Material Criticality. *Resources, Conservation and Recycling, 155*, 104617. https://doi.org/10.1016/j.resconrec.2019.104617

Sovacool, B. B. K., Ali, S. H., Bazilian, M., Radley, B., Nemery, B., Okatz, J., & Mulvaney, D. (2020). Policy Coordination is Needed for Global Supply Chains. *Science*, *367*(6473), 30–33.

Umbach, F. (2021). Critical Raw Materials: Assessing EU Vulnerabilities. Gisreportonline, https://www.gisreportsonline.com/r/critical-raw-materials/

United Nations. (2021). COP26 The Glasgow Climate Pact. In *UN Climate Change Conference UK 2021*. https://ukcop26.org/the-conference/cop26-outcomes

United Nations Environment Programme International Resource Panel. (2020). Resource Efficiency and Climate Change: Material Efficiency Strategies for a Low-Carbon Future. In *International Resource Panel*. https://doi.org/10.5281/zenodo.3542680

United States Department of Energy. (2011). *Critical Materials Strategy*.

United States Geological Survey. (2022a). *2022 Final List of Critical Minerals*. https://www.usgs.gov/news/national-news-release/us-geological-survey-releases-2022-list-critical-minerals

United States Geological Survey. (2022b). Mineral Commodity Summaries 2022. https://doi.org/10.3133/mcs2022

Vidal, O., Goffé, B., & Arndt, N. (2013). Metals for a Low-Carbon Society. *Nature Geoscience*, *6*(11), 894–896. https://doi.org/10.1038/ngeo1993

Watari, T., Nansai, K., & Nakajima, K. (2020). Review of Critical Metal Dynamics to 2050 for 48 elements. *Resources, Conservation and Recycling, 155*, 104669. https://doi.org/10.1016/j.resconrec.2019.104669

Wiedmann, T. O., Schandl, H., Lenzen, M., Moran, D., Suh, S., West, J., & Kanemoto, K. (2015). The Material Footprint of Nations. *Proceedings of the National Academy of Sciences of the United States of America, 112*(20), 6271–6276. https://doi.org/10.1073/pnas.1220362110

Wilburn, D. R. (2012). Byproduct Metals and Rare-Earth Elements Used in the Production of Light-Emitting Diodes—Overview of Principal Sources of Supply and Material Requirements for Selected Markets. In *Scientific Investigation Report 2012–2015*. https://pubs.usgs.gov; http://pubs.usgs.gov/sir/2012/5215/

Wilson, F. (2020). Strategies for Securing Critical Material Value Chains. Perth USAsia Centre, April 2020. https://perthusasia.edu.au/

Wood Mackenizie (2022). *Global Rare Earths Short-term Outlook*. April 2022 Report. https://www.woodmac.com/reports/metals-global-rare-earths-short-term-outlook-april-2022-150025745

World Bank. (2017). The Growing Role of Minerals and Metals for a Low Carbon Future. In *World Bank Publications*. https://doi.org/10.1596/28312

Yang, Y., Walton, A., Sheridan, R., Güth, K., Gauß, R., Gutfleisch, O., . . . & Binnemans, K. (2017). REE Recovery from End-of-Life NdFeB Permanent Magnet Scrap: A Critical Review. *Journal of Sustainable Metallurgy*, 3(1), 122–149.

Zepf, V., Simmons, J., Reller, A., Ashfield, M., & Rennie, C. (2014). *Materials Critical to the Energy Industry. An Introduction. 2nd edition.* **[AQ1]**

Chapter Two

Metaphors and Expectations for the Realization of 'Zero Carbon' Steel Production

Annika Skoglund

Tipping points, a metaphor that commonly is used to describe critical thresholds (O'Riordan and Lenton 2013), like major shifts in biophysical systems (McKay et al. 2022), has increasingly entered the vocabulary of economic experts to speed up and realize decarbonization at scale (Systemiq 2023). While tipping points in nature refer to negative losses, such as 'the loss of the Amazon rainforest or the West Antarctic ice sheet' (Lenton et al. 2019:592), the same metaphor used in the economic system commonly alludes to positive gains of growth (e.g., see Brunner 2008). In nature, tipping points are apprehended negatively as leading to unwanted irreversible changes (Lenton et al. 2019), whereas in economic systems, they are often apprehended positively as resulting in creative destruction and desired alternatives (Systemiq, 2023). As a metaphor, a tipping point is a 'means for characterizing, and giving shape to, uncertainty, interdependence, turbulence, and crises, and for unleashing the creative power of imagining' (O'Riordan, Lenton, and Christie 2013:6). Tipping points call for critical thresholds that per definition cannot be causally controlled—but continuously shaped. The metaphor can thus be invoked strategically from a processual perspective, to mobilize and manoeuvre an emerging future full of uncertainty.

In stark contrast to such a strategic processual perspective, calculation of cause and effect for rational decision-making is often used in efforts to initiate fundamental technological transformations and tipping points, as is observable in the policy report titled *The Breakthrough Effect—How to Trigger a Cascade of Tipping Points to Accelerate the Net Zero Transition*

(Systemiq 2023). In its genre of formal forecasting, the report keeps close to efforts of controlling the future, stipulating that market tipping points can be 'triggered' stepwise by facilitation of a limited set of identified cost-based conditions. The aim is to causally facilitate fast adoption of new technological solutions that feed exponential growth and rapid replacement of old solutions. Targeting specific tipping points is conceived as key to this generation of speed, and as presented at the World Economic Forum meeting in Davos in January 2023, decarbonization is even suggested to be accelerated via 'super-leverage-points' (Systemiq 2023). Directing actions systematically towards their triggering conditions is judged to be the most efficient way forward, since these will 'catalyse tipping cascades across sectors, to make the Breakthrough Agenda vision a reality' (Systemiq 2023:16). Nicholas Stern, a leading Professor of Economics, thus suggests that a better understanding of the rational conditions for tipping points is crucial for 'unleash[ing] accelerated growth and the new sources of prosperity' (Stern 2023). In other words, tipping points are treated as rational future-making devices that should be targeted by investors to get 'bang for your buck,' according to Gail Whiteman, an influential Professor of Sustainability (Whiteman 2023). Formal forecasting of tipping points is thus provided to economic actors as a sound basis for taking rational decisions in correct directions of certain technology paths.

Neither the tipping point metaphor, nor the breakthrough one, are in the call for super-leverage-points and triggering of cascades (Systemiq 2023) explicitly used to strategically elicit creativity and imagination (O'Riordan, Lenton, and Christie 2013). Still, what previous research on forecasting has shown is that "formal articulation of futures takes place in situations where expectations abound and informal assessments are continuously made. In other words, foresight necessarily occurs in a 'sea' of expectations" (van Lente 2012:777). The rational calculations produced in efforts to secure some order in the desired disruptiveness and beneficial arrival of a future 'Breakthrough Effect' (Systemiq 2023), conflates with informal and incalculable expectations (van Lente 2012, Borup et al. 2006). This means that neither policy makers, nor economic actors, can rely solely on rational expectations theory and assumed (equilibrium) points for when 'tipping' beneficially would occur (Beckert and Bronk 2018).

The limitation of calculations and formal forecasting is particularly prevalent in the case of technological and social transformations, since these make economic actors increasingly exposed to an unruly growth of high expectations, with risks of disappointment (Ruef and Markard 2010). In addition to all the calculation and rational decision-making that fill reports akin to *The Breakthrough Effect*, organisations with ambition to transform therefore need complementary methods of less formal foresight that facilitate creative choices that are intellectually unconfined to rational decision-making

(Beckert and Bronk 2018). This has, for example, been discussed in relation to corporate vision statements that seek to 'future-proof' business (Hazard 2011), and in relation to entrepreneurial constructions of 'opportunities' for the making of future goods and services (Berglund and Dimov 2023). What these and other studies clarify is that expectations can be strategically used by corporate actors and policymakers as a future-making device, to foster legitimation, collaboration and direction for the realization of certain technology paths (van Lente 2012). Consequently, if the ambition is to accomplish a technological transformation, for instance metaphorically espoused to play out as a market tipping point, the way in which visions[1] and expectations of such are expressed also provide an important condition for action and realization of such a future—in the present.

To complement the existing focus on formal forecasting for the triggering of tipping points, this chapter will address the ignored role of how metaphors generate expectations for the technological transformation and decarbonization of a specific industry, steel production in Sweden. Energy transitions have continuously impacted Swedish steel production processes (Johansson 2014), including a shift from charcoal to coking coal, another shift from oil to natural gas, and current investigations of a shift to biomass (Nwachukwu et al. 2022) and hydrogen (Tang et al. 2021). Even with emission related improvements, from energy efficiency to steel recycling efforts (Berlin, Feldmann, and Nuur 2022), the mass production of advanced steel grades in Sweden nevertheless amounts to about 5.5 Mton CO_2 equivalents per year (Fossil Free Sweden 2023), representing approximately 10% of the yearly national emissions (Vogl, Rootzén, and Svensson 2019). Steel is thus of particular interest due to how 'the past' of steel production has been identified to need a 'breakthrough' to properly disrupt its fossil fuel lock-in (Systemiq 2023). From general improvements of recycling and increased use of scrap and material efficiency, future steel production is even suggested to require implementation of immature technologies and a portfolio of solutions depending on cost-competitiveness in particular geographical locations (Systemiq 2023). If tipping in one place, a worldwide transformation is assumed to unravel. This should be of particular interest to the Swedish steel industry and its association, Jernkontoret (the 'Iron office') (Jernkontoret 2017), who share the vision that Sweden should take lead in the steel transformation.

Jernkontoret has approximately 20 members and makes their mutual need for a clear vision explicit in a Klimatfärdplan (Climatejourneyplan) (Jernkontoret 2018). Since 2011, the association has developed its 2050 vision statement, *Steel makes a better future*, in close dialogue with stakeholders (Jernkontoret 2015, 2017, 2018). Of interest here is the long-term thinking that the association facilitates, at the same time as their vision statements are grounded in an explicit processual perspective. Jernkontoret focuses on 'a vision on the way to be realized' (Jernkontoret 2017) and not so much a future to come. The future is conceived to be continuously shaped, in the present, through the production of decarbonized advanced steel grades in Sweden. Furthermore, the question is how Jernkontoret's processual vision statements, which nurture long-term thinking about a continuous shaping of the future, feeds into other expectations of breakthroughs within the Swedish steel industry? How may metaphors such as tipping points, breakthroughs and leaps infuse expectations of technology-based threshold effects for the realization of a deep decarbonization of steel production?

To investigate these questions further, the chapter first introduces how the technological transformation of the steel industry has been calculated upon, and rationally envisioned, in relation to tipping points and a desired energy transition. Then follows a theory section on the sociology of expectations, explaining how this literature has been developed for a better understanding of technological and social transformations. Metaphors that elicit uncertainty and expectations about the future of steel are thereafter analysed in the conflation of Swedish industrial and climate policy, whereafter conclusions are drawn. As a complement to existing knowledge about rational decision making for the realization of transformations, the chapter explains how expectations can be mobilized during the realization of specific technological 'breakthroughs,' in this case a stepwise accomplishment of nearly 'zero carbon' steel production via hydrogen.

TRANSFORMING THE STEEL INDUSTRY

The Breakthrough Effect report focuses its forecasting on ten main sector tipping points to be accelerated, of which one is steel production. For example, the European Union Carbon Border Adjustment Mechanism (CBAM) is suggested to be part of the triggering of tipping points (Systemiq 2023). The CBAM will be applied to importers of iron, steel, cement, artificial fertilisers, electricity, as well as hydrogen. The aim is to phase out the use of free allowances in the EU Emissions Trading System by 2026, and set a price on the CO_2 emitted during production processes (European Commission 2023).[2] Forecasting based on the CBAM, could according to the report pave the way for one of the new technologies for fossil-free steel production, HYBRIT (Hydrogen Breakthrough Ironmaking Technology) (Systemiq 2023:46), first implemented at pilot scale in north Sweden. This technology is dependent on the development of fossil-free (sometimes referred to as 'green' or 'renewable') hydrogen, produced with electricity generated without CO_2 emissions. This is also partly why north Sweden has been identified as a suitable location for starting the call for global transformation of the steel industry (Pei et al. 2020, Öhman, Karakaya, and Urban 2022).

The possibility to use hydrogen for steelmaking has in theory been well-known, and is in that sense not a new revolutionary scientific insight (Brännberg Fogelström, 2022). Hydrogen produced without CO_2 emissions would however not only be useful for DRI (direct reduced iron-based) steel making[3], but also for the decarbonization of fertilizer production, aviation and long-haul shipping (Systemiq 2023). Hydrogen is thus conceived as a general enabler of tipping points within several industries, an aspect that should be taken into consideration, according to the authors of *The Breakthrough Effect* report. The current pilot stage of green hydrogen is identified to need further facilitation, to be realized at commercial scale, which in turn would mobilize important conditions for industry tipping points to be triggered across industries (Systemiq 2023). By starting in a few places with green hydrogen for steel, industrial transformations are assumed to be activated broadly, causing 'tipping cascades' that are deemed necessary for a realization of 'the Breakthrough Agenda' (Systemiq 2023:16).

One condition identified for the competitiveness of green hydrogen for steel is the cost of electrolyser (IRENA 2020). Another calculation stipulates that the cost per ton of steel produced, when using green hydrogen DRI, would need to be less than conventionally produced steel in a basic oxygen blast furnace. This competitiveness of the new technology is currently calculated to be accomplished when the carbon pricing of steel is about $100/ tCO_2, and when the green hydrogen price is about $1.2–2.2/kg H_2 (Systemiq 2023:70). The ambition is thus to secure these conditions for rational-decision making in different locations to facilitate a threshold effect—a breakthrough or tipping point—and thereby accomplish a realization of fossil-free steel production at commercial scale in practice. According to the non-profit industry coalition called 'We mean business' (2022:18), capital flows should furthermore be directed towards those companies that are focused on alignment with net-zero pathways, to 'accelerate these shifts.' Overall, investments to secure the growing demand for steel 30 years from now are estimated to need $47 billion globally (We mean business 2022). And if the steel industry is to transform and become net-zero-compliant, the current calculations stipulate that new technologies will need an additional annual investment of about $8–$11 billion (We mean business 2022). For example, the European Commission has approved state aid measures of $2.85 billion to ArcelorMittal (France) and Thyssenkrupp (Germany), to advance steel production towards the use of green hydrogen (We mean business coalition 2023).

Climate policy is in these rational economic ways becoming increasingly embedded in industrial policy, and several actors in the steel industry have been positively affected by the new EU regulations, receiving investment support to become part of the tipping point facilitation. In north Sweden, the future investments are estimated to total around 100 million USD, encompassing both the production of green hydrogen and new fossil-free steel (SCPC 2023a). Major investments have also already been made at an early stage to comply with the Paris climate targets, attracting both Swedish and international companies to the region. In 2021, the Swedish steel corporation SSAB[4] produced the first batch of fossil-free steel in the world. This was possible at a demonstration stage through the HYBRIT project, which is a research and innovation collaboration between the Swedish state-owned energy utility Vattenfall, the mining corporation LKAB[5], and SSAB (Pei et al. 2020). LKAB, is a public limited company, fully owned by the Swedish state, and their mines are mainly located in the 'ore fields' in north Sweden. As a major shareholder of SSAB, LKAB wishes to secure the transformation towards fossil-free steel. LKAB also suggests that a 'global transformation' is underway, which is 'the greatest challenge of our time' (LKAB 2023b).

The commercial advancement of fossil-free steel in Sweden is additionally not limited to HYBRIT and its industrial partners. Another example of how the steel industry is working actively with an expected energy transition, or tipping-over into hydrogen, is OVAKO (part of the Nippon Steel Corporation). OVAKO mainly recycles steel scrap and uses electrical arc furnaces but has also invested in a new hydrogen plant south of Stockholm to replace liquified petroleum gas (LPG) used in the heat treatment furnaces (Excell 2023). A large global technology company within mobility, ZF, has also signed a 7-year agreement with the Swedish start-up H_2 Green Steel (2023). ZF, which has 168 production locations, will reduce 2.3 million tonnes of CO_2 emissions in their production by purchasing steel produced by low CO_2 emissions. During the binding agreement with ZF, H_2 Green Steel has an additional ambition to engage with ZF's sub-suppliers, to drive reduction of carbon emissions along the whole value chain (H2 green steel 2023). SSAB, on the other hand, will partner with the BE Group, a steel service company active in Sweden, Finland, the Baltic States and Poland. By working together over a longer time period, SSAB and BE Group aim to guide BE's customers (SSAB's end users) towards new market positions based on green premium end products (SSAB 2023). There is thus a considerable commercial momentum for deep decarbonization of steel in north Europe, where competition is not only developing in relation to conventional steel production stuck in 'the past' of fossil fuels, but between old and new companies that are focussed on 'the future' of hydrogen.

THE SOCIOLOGY OF EXPECTATIONS AND TRANSFORMATIONS

Expressed or kept in silence, expectations do not only play an important role in everyday interactions and hopeful rhetoric, common examples being the promissory but fragile 'American dream' or 'Chinese dream.' Expectations are also germane to areas saturated with future-oriented thinking, such as technological development (Borup et al. 2006), economic decision-making (Beckert 2013) and entrepreneurship (Berglund and Dimov 2023). Demands for innovativeness and national competitiveness can even give rise to what Konrad (2006) elaborates on as 'hyped' or 'high rising' expectations (see also Ruef and Markard 2010). Expectations about technological and economic advancement are in these cases reciprocally accentuated, underwritten by taken for granted requirements of material progress. With calls for transformations of entire industries and societies, such a notion of progress has nevertheless been reformed, supported by an even more powerful accentuation of expectations of a better future to be made. Positive projections of a sustainable future have bolstered high expectations of continued technological and economic progress.

Expectations of transformations towards sustainability are also different from expectations of sustainable transitions. A transformation requires a fundamental technological leap and engagement with the unknown that incremental innovations do not offer. If a transition slowly builds on the past to incrementally change the future in the present, a transformation relates differently to the past to guide future-making practices, that is, practices that attempt to curtail the future (cf. Wenzel et al. 2020). Conceptually, a 'transformation' is understood to disrupt the past in attempts to create a break with the present status-quo or business-as-usual, and thereby accomplish something more novel, perhaps radically new and unknown (e.g., see Stirling 2015). It is a break with the past that feeds new methods for forward-looking that are less dependent on the known past, and more dependent on the unknown future, and expectations of such. Hence, a transformation requires incalculable expectations that are open to what is sometimes also referred to as 'unknown unknowns' (cf. Porter, Kuhn, and Nerlich 2018:881). Rather than relating analytically to these unknowns as manufacturing of 'the future' as radically unknown, expectations of such can be traced in relation to their everyday affirmation of uncertainty. That is, expectations during transformations do not seek to directly control the uncertainty they offer. Instead, they produce and affirm uncertainty in the present to orient action and accelerate it, particularly if abstract images of 'unknown unknowns' are elicited through metaphors such as tipping points and breakthroughs.

Expectations about the future are thus accentuated, and particularly pronounced, when a technological transformation, 'breakthrough' or 'tipping point,' is called for. One reason may be that transformative agendas unsettle any profound reliance on the past, which otherwise is common in formal foresight calculations. In expectations of a technological transformation, the past is rhetorically drawn on as an adversary to a better-looking future. Expectations are during calls for transformations seeking a future properly detached from the past, which is disqualified as a yardstick, inept to use as a basis for the future. While some scholars would prefer to use concepts closer to imagination (Komporozos-Athanasiou and Fotaki 2015), imaginaries (Beckert and Bronk 2018) and imagined futures (Oomen et al. 2022), to convey such a detachment, expectations are in this chapter conceptually assumed to be closer to the everyday future-making of organisations when these are to *realize* images and visions of a technological transformation (cf. Pettit, Balogun, and Bennett 2023). Even if both imaginaries and expectations have been conceptualized as performative, there is an analytical difference. To simplify, expectations are 'earthy' and embedded in transformative practice, while imaginaries are 'lofty' and embedded in transformative rhetoric. Images of the future that are wild and impractical (Tutton 2017), as well as imagined possibilities and alternatives 'far away,' are less easy to grasp and research (cf. Bell and Mau 1973). In comparison, expectations are in this chapter conceived to be less about fictions and fantasies, and more about an expressed inevitability and enactment of a certain desired future that already is recognized as emergent.

Hand in hand with improved data crunching and even forecasting that is less reliant on historical data from 'the past,' the organisational 'steering' role of expectations is fundamental for transformations. Forecasting models are not necessarily used for rational decision-making but can be mobilized as a philosophical device to facilitate reflection upon one's actions as part of a collective steering into the future (Brown et al. 2004). Expectations are thereby recognized for providing extra motivation, nurturing willingness to take action in directions that affirm uncertainty and largely unknown futures. Expressed expectations that are used to collectively envision a transformation, such as the metaphors of tipping point or breakthrough, are conceived to have certain effects. For example, if market tipping points are desired, and natural tipping points undesired, expectations about these possible polarized futures will have direct consequences for the present enactment of the preferred tipping point. Without guarantees for a successful realization of the desired future (Konrad 2006, Ruef and Markard 2010), expectations are assumed to bring about the creation of legitimate action, collaboration and direction (van Lente 2012).

Expectations, analysed as statements about possible futures, are not only conceived to describe the content of a vision—the improved future that potentially will come at a later stage. The expressed expectations are in addition treated to have a direct effect in the here and now, why 'the future' is studied as present enactments of it (van Lente 2012). However, an analysis of what expectations about the future performs, in the present, needs to consider many different things, depending on the context. For example, the person who makes a bold statement about a desired future they wish to be part of, creates both a position for themselves as being anticipatory ambitious, as well as momentum towards the future they express. In turn, this can feed further expectations in the same direction, if the vision of that future also is desired by others, who accommodate the expectations. Others may start to expect that the person who made the bold statement is actually very ambitious and committed to a certain future, and thereby continue to support such in a way that enforces it. And if the assumedly bold, ambitious, committed person would not live up to the growing expectations, perhaps further support would be provided to rescue the expected. This could not only save the by now acclaimed ambitious person from failure, but also save those who accommodated the expectations, and perhaps even promoted the anticipated expectation-rich person along the way. Instead of using rational decision-making based on information given about experiences, competences and formal skills in a CV (Curriculum Vitae), expectations expressed by someone and accommodated by others, can thus lead to investments and a clarification of duties (Borup et al. 2006). This is how shared expectations are able to set in motion a collective realization of the future in the present.

Other types of self-fulfilling prophecies can also occur. For example, if a certain future is produced through calculated predictions, the figures may be rescued along the way if the prediction is under threat to fail (van Lente 2012). This is called Moore's law and clarifies how rational decision-making is supported by expectations of such to be correct, used as a yardstick for progress. Calculations can thus be made to work due to the expectation that something rationally calculated upon should work as calculated. Instead of acting as a positivistic proof of an accurate prediction of 'reality,' the calculated future is thus realized through expectations. Van Lente (2012) illustrates how Moore's law is strategically used by businesses when making investment decisions on different R&D (Research and Development) goals. Businesses assume that expectations will feed self-preservation. That is, 'when the promised specifications run the risk of not being met, additional measures are needed, such as entering into strategic alliances' to build momentum around the expectation (van Lente 2012:773). During policy calls for tipping points and production of increased uncertainty about the future, it is thus expected that corporations accentuate their partner relations to secure an accumulation of shared expectations.

There are nevertheless no guarantees, since the way in which expectations about the future influence present future-making practices is non-predictable (Ruef and Markard 2010, Konrad 2006). Many expectations about the future are diffused along the way, getting detached from a certain future that was more clearly expected in previous statements. Despite having many followers, rational foresight that made believable promises about a certain future's desirability can also prove to be impossible to invest further expectations in. High expectations can quickly turn into low expectations when promises are not fulfilled (Ruef and Markard 2010). However, what is hoped for in cases where expectations are used strategically, is that enough people will expect the same thing and feel required to keep investing in the expected direction of the future. It is the expectation that something should, rather than could, happen, that leads to strategic direction and a potential realization of transformations within business.

For example, expectations may be used to set priorities and legitimize certain futures, while disregarding other alternatives. In these cases, expectations are statements that make specific *claims on* the future, wishing it to contain certain aspects and thereby to be processed in a curtailed direction, or what Jenny Andersson (2018:83) explores as the playing out of 'future stakes and interests.' This means that dominant images of the future develop due to established interests, or normative preferences, that in turn also are indebted to expectations (Andersson 2018). In the case of fossil-free steel, economic interest and rational calculations would thus have an effect on the informal expectations and visions of the future of steel, and the other way around. Informal expectations guide the making of formal forecasting of steel futures, and the process of formal and informal foresight is inseparable, yet possible to explore analytically.

An analysis of silent expectations about the future, or the 'inhabitation in the mind of an imagined future state of the world' (Beckert 2013:220) is difficult to pursue. The risk is that the researcher takes an already completed, observable act, and interprets it as a proof of a previous expectation. Instead, it is crucial to stay with the expectation as temporally embedded, seen as pointing to an emerging projected possible future (cf. Mische 2009). How that is done in research practice is nevertheless not clear. Beckert (2013:220), for example, studies 'fictional expectations' by turning empirically to 'narrative form as stories, theories, and discourses,' after which analytical attention can be given to how the research participants consider their actions to be taken in the direction of futures to come (cf. Mische 2009). Expectations are thus theorized upon as inextricably coupled to why certain actions are initiated—or to paraphrase Beckert (2013:220)—expectations about possible futures is an important element for understanding and explaining present action. Furthermore, in the particular case of 'fictional expectations,' these are treated as pointers to the 'non-observable,' where anticipatory expressions 'pretend' that the future will consist of certain things that 'may or may not materialize' (Beckert 2013:224). Analytically, focus can thus only be what such expressions accomplish in the present, with sensitivity towards how expectations suspend disbelief by designing fantasies that are possible to commit to practically (Beckert 2013). That is, even if many are not materialized, expectations in technological and economic 'futuring' are continuously calibrated in relation to their practical feasibility.

GROWING EXPECTATIONS ABOUT THE FUTURE OF STEEL

Steel is one of the most dominant materials in our everyday lives, and its past contribution to modernization is often brought up when changes in the industry are discussed (AFRY Management Consulting 2022). The immense objects and applications resulting from sedimented human efforts and pride contribute to path-dependence. The steel industry has, for example, been characterized as reluctant to change due to huge sunk capital investments, explanatory for why new research and development activities historically have been limited (Luiten, van Lente, and Blok 2006). The different steel corporations have invested in similar production processes (Luiten, van Lente, and Blok 2006), making it practically difficult for any single actor to take the first transformatory step. It is thus understandable that the industry itself, and scholars with an interest in it, mainly have attended to the past, rather than the future, in explanations of present actions taken.

Historical developments within the Swedish steel industry have been richly depicted from technological, economic and cultural perspectives (Jonsson 2009, Karlsson 2016:1080, Jakobsson 2009). Jansson (2023) portrays how, since at least the seventeenth century, steel production has been advanced by a long accumulation of expert knowledge, underpinned by state support that facilitated strong inter-industrial relations. First acting as a logistical network of companies, the Swedish steel industry was later organised in relatively stable cartels. In the 1950s different steel producers were engaged in numerous cartels, forming an increasingly 'complicated network' (Karlsson 2016:1080). Characteristic of the remnants of this network is the lack of direct competitors, since the different producers historically have complemented each other and specialized in different steel grades and products. The industry's prosperous historical legacy of collaboration is also reflected in the steel industry association Jernkontoret, founded in 1747. In comparison to the perceived lock-in effects and barriers to change within the steel industry globally, Swedish strong inter-industrial relations have opened up for a willingness to change. Instead of demanding any single actor to take the first transformatory step alone, Jernkontoret gathers the industry under the mutual vision that Sweden should take a lead in the transformation.

Jernkontoret's (2017) overarching vision for the collective industry called, '*Steel makes a better future*,' is not likely to become irrelevant or inaccurate along the journey towards the final deadline, 2050. The vision is expressed from a processual perspective (Jernkontoret 2015), much akin to how sociologists with an interest in future studies early saw their own academic task to be a directly enacted betterment of an emerging future (Bell and Mau 1971). Jernkontoret's vision embeds the future within steel as a material, and key to eliciting betterment is steel itself. There is nothing wild or fictional involved in their overall statements. They do not question 'what the future will bring,' nor what it 'will be like' (Brown et al. 2004:9) but stay with the loose notion of 'betterment' conceived as incremental techno-optimistic progress. To assure that the steel produced actually will make a positive contribution of betterment, however, Jernkontoret also presents three core areas that need strategic attention. Notably, their three visionary pillars facilitate expectation and anticipation, not for thinking *about* the future—but to foster preferable action *into* the future (Brown et al. 2004).

The first vision statement revolves around technology, innovation and research, stating that Swedish steel industry will be vanguard and lead the technological development in the world (Jernkontoret 2017). The advancement of technology is not expressed in revolutionary terms of 'breakthrough' or 'tipping point,' whereby the association is not 'breaking' with its long history. On the contrary, the association is careful not to exclude the competent past, in provision of consistency to the long Swedish superiority exemplifying a continuous growing breadth and depth of already existing edgy solutions presented by their members (Jernkontoret 2015, 2017). The second visionary area, in contrast, relates to the human capacity to think differently about technology and even move 'beyond' what so far has been expected in relation to the past of steel. The vision statement morphs into less controllable aspects of incalculable and fictional expectations, exemplified through art in the initiative 'Beyond Steel,' (Jernkontoret 2023b). A documentary elaborates on a more contemporaneous image of steel, from crafting of knives to origami folding and a steel scented perfume that captures abstract steel sentiments (Jernkontoret 2023a).

The art initiative stretches to Jernkontoret's vision of a creative work environment that should augment boundaryless thinking, unconfined to rational decision-making (Jernkontoret 2017). This call for boundaryless thinking partly aligns with studies that point to how organisations with an ambition to transform have to focus less on formal foresight, to rather facilitate creative choices that are intellectually unconfined (Beckert and Bronk 2018). Disruption and uncertainty are consequently affirmed, and the ongoing transformation is by Jernkontoret clearly expected to happen through an accentuation of individual creative efforts, and employees who think hard about their continuous shaping of the future. The second pillar of Jernkontoret's vision thus complements the first one, which was expressed through a more deterministic view of accumulated technological knowledge that develops stepwise, path-dependently. To keep Sweden technologically competitive, expectations are thus de-centred to mainly grow within organisations and their prosperous work environments, and within individuals and their boundaryless thinking.

The third visionary aspect extends the focus on organisational and individual transformatory ingenuity to an expectation of increased interconnectedness. The environment is brought in for calibration of a legitimate direction for the emerging future (van Lente 2012). Jernkontoret stipulates that only socially valuable products are to be commercialized, and those that threaten the environment are not socially valuable. To conclude, Jernkontoret clarifies the interrelations between these three areas, technology/human/environment, and makes them mutually dependent. It is only with leading technological development that environmental value can be secured, and it is only with a stimulating work environment that employees will engage in vanguard R&D. Lastly, it is only by supporting socially valuable products and societies that the most competent and skilled people will be attracted to work with the innovation of steel (Jernkontoret 2017). As a vision, the three pillars are seemingly operable by many different actors, at the same time as Jernkontoret pushes thinking closer to reflections about 'the future' without much disturbance of the strong connection to the past. The vision itself may even become a device for reflexivity, making the ongoing future-making within the industry more explicit to both Jernkontoret's members and stakeholders.

Since at least the mid-2010s the steel industry's interest in the past has thus been complemented with, and sometimes replaced by, a greater interest in the future. The barriers connected to 'the past' are consequently conceived to be less rigid, loosened up both in work and thought, generating high expectations of a steel industry transformation to come. Some point to the end-of life of past investments, and the rational need to shift current processes for the sake of both inevitable technological renewal and future desired decarbonization (Kushnir et al. 2020). Due to the already established environmental interest within the Swedish steel industry (Jansson and Ryden 2023, Sandberg et al. 2001), Swedish actors and stakeholders are also conceived to be prepared to meet the new demands (see however divergent environmental interests detailed in Alriksson 2013). 'The future' has become more practically feasible (Beckert 2013) to think hard about and work towards in the present. This growing expectation of a transformation of the steel industry reflect the Swedish government aim from 2021, to reach net-zero greenhouse gas emissions by 2045, whereafter negative emissions should be obtained (SCPC 2023a). A binding target for Sweden has also been set, to increase net removals of about 4 million tonnes annually by 2030, in comparison to the average emissions between 2016–2018 (SCPC 2023a). The government in 2021 considered it to be important for Sweden to 'show international climate leadership,' undertaking emission reductions that 'far exceed the requirements under the EU Effort Sharing Regulation' (Government Offices of Sweden 2021). The previous government thus set the expectations high, for industrial actors, with ambition to take lead globally (Brodén Gyberg and Lövbrand 2022). However, this ambition was not shared by the next government, according to the Swedish Climate Policy Council (SCPC), which is an independent, interdisciplinary body of experts that evaluate alignment with the Paris climate targets (SCPC 2023b).

In relation to previously expressed expectations, the SCPC (2023c) proclaims that Sweden has lost pace and should work harder to fulfil the expectations of an accelerated decarbonization. The first part of the annual report 2023 further explains the difference in policy orientation, and divergent expectations about the future, held by the old and newly elected government. The second part of the report also takes the evaluation of the new government further, pointing to the already evident meeting of shared expectations in the realization of an 'ongoing large-scale transformation of industries in Norrbotten and Västerbotten' (SCPC 2023a:3). In their words, '[p]hasing out all fossil fuels and reducing greenhouse gas emissions to zero entails a fundamental transformation of society' (SCPC 2023a:10). Furthermore, the phasing out is not emphasizing the new Emission Allowances and other flexible mechanisms for exchange of emissions between EU countries, as laid out in the EU report *The Breakthrough Effect* (Systemiq 2023). Instead of relying on market mechanisms and exchange of allowances between countries, the main focus is the technological transformation needed to accomplish a 'real' reduction of emissions. The SCPC further reflects on how the EU is picking up on the previous, more ambitious climate goals that were held by the former Swedish government. Since Swedish industry already has been working towards the higher expectations, the SCPC concludes that it will be easier to comply with the improved future EU goals.

Aligning with this expectation of a true material reduction of CO_2, the steel industry has also shifted its decarbonization vocabulary. Instead of speaking about 'net zero' or 'carbon neutral,' the notion of 'zero carbon' is growing in popularity. LKAB (2023a), for example, states that in 2045, they 'will be down to zero carbon dioxide emissions from [their] products and processes.'

This brings lofty images about the future 'down' to expectations of a transformation at the operations of the emitter, who can work with decarbonization in a more hands-on everyday way, stepwise towards 'near' zero and a specified deadline. In comparison to the popular notion of 'fossil-free,' 'zero carbon' is not as abstract, but blends rational calculation with informal expectations of such a figure as 'zero' to become accurate in the future (van Lente 2012). The vision is thus rationally grounded in a number and linear time, with possibility to work practically with expectations of increasingly detailed calculations of 'almost' zero emissions, rather than inaccurate and lofty promises of totally 'fossil-free' processes and products. The shared expectation of a possible 'zero' is thus formative of certain technological solutions, which steers into 'real' or 'deep' decarbonization.

In comparison to the processual vision statement of Jernkontoret (2015), the end goal of 'zero carbon' is not representable as a transition but expressed to need a major revolutionary or radical technological shift that 're-invents' the future. SSAB (2022b), for example, expresses in their annual report that they aim to 'reinvent the future of steel.' They are also an early mover, who expressed their expectations about a transformation before such was legitimately spoken about internationally (De La Garza 2022). Even if SSAB since 2006 has worked closely with the vision 'A stronger, lighter and more sustainable world,' the new technical development of HYBRIT can be considered a break with the old methods of steel production. SSAB thus specifies their expectations about a transformation, stating that their 'upcoming breakthrough fossil free technology is the first true steelmaking transformation for centuries . . . adapted to the future' (SSAB 2022a:2). Eliciting expectations of a breakthrough to come, this statement echoes *The Breakthrough Effect* report (Systemiq 2023), and adds to the mobilization of other threshold metaphors, such as tipping-points, which welcome uncertainty as productive for further creative thought (O'Riordan, Lenton, and Christie 2013).

A threshold effect is also alluded to in funding programs such as the Industrial Leap (Swedish Energy Agency 2023). The Swedish Energy Agency, responsible for the program, pushes the expectations about a transformation even beyond 'zero,' calling for technological leaps that are not only expected to reduce process-related emissions, but achieve negative emissions. Similarly, Professor Johan Rockström, Director of the Potsdam Institute for Climate Impact Research, emphasizes the need of such a move towards negative carbon numbers, by eliciting positive market tipping points (Rockström 2023). According to him, when ten to twenty percent of the market is on board, it is possible to 'cross the valley of death,' tip over and have a positive effect on the majority of the market (ibid.). In comparison to the metaphor of tipping point, a breakthrough nevertheless invests more in promises about the future, for example a certain technology. And it is towards this technology, which works as a container for the breakthrough metaphor, that expectations and actions are directed 'as if' the future solution, in fact, will deliver on the promised threshold effect in a positive way. As a metaphor, a breakthrough is thus not only shaping uncertainties and turbulence to better manage such states within technology development and business (O'Riordan, Lenton, and Christie 2013). Expectations of a breakthrough can also park potential disbelief (cf. Beckert 2013) about the ongoing technological transformation by pointing to a specific promissory technology that becomes a requirement.

CONCLUSION

The past has played an important role in the steel industry, and the Swedish steel association, Jernkontoret, stays loyal to this legacy by making vision statements from a processual perspective (cf. Bell and Mau 1971). Steel has always and is continuing to 'shape a better future' (Jernkontoret 2017) in a 'vision on the way to be realized' (Jernkontoret 2015). Instead of eliciting expectations about a certain future 'far away' that is impractical (Tutton 2017), Jernkontoret steers action into the future (Brown et al. 2004) and makes their members aware of their own already existing future-making practices. At the same time as accumulated knowledge about steel production is acknowledged as important, Jernkontoret also points towards less rational decision-making in favour of boundaryless creativity and value-based interconnectedness. And it is perhaps here that expectations of, for and during a transformation become more visible. Instead of expecting specifics themselves, Jernkontoret de-centres the work of expecting, and future-making, to happen among individual employees who should bring Sweden into a leading position for the realization of 'zero carbon' steel production.

As a complement to Jernkontoret's processual approach with focus on individual ingenuity for future-making, metaphors such as breakthrough, tipping point and leap give rise to expectations of uncertainty and a transformation to come. The past is no longer as important, but is referred to as a counter figure, something to work against and disrupt, for the realization of a more fundamental transformation of the present. Despite metaphorical images of a turbulent and uncertain future (O'Riordan, Lenton, and Christie 2013:6), hopeful expectations are elicited, supported by expressions of an already ongoing deep decarbonization that is desired and inevitable. The future of Swedish steel is emerging through expectations of almost 'zero carbon' to be realized, stepwise, until 2045 or 2050. Where the number 'zero,' in comparison to the notion of being 'neutral' or carbon 'free,' brings the expectations down to everyday activities that are practically feasible, to be calculated upon. It positions the expectation of decarbonization to happen at the operations of a certain CO_2 emitter, who has to 'meet expectations' shared and negotiated among a network of actors and stakeholders. Disbelief is thus suspended through alliances (cf. Beckert 2013), driving expectations of mutual investments in more specified promissory technologies (van Lente 2012), such as 'green' hydrogen, that is turned into a requirement. To conclude, the call for tipping points and breakthroughs (Systemiq 2023) accentuates expectations that express a clear disconnection from 'the past,' which does not seek to control a specific future, but steers into 'the future'—left unspecified yet remains believable and actionable.

ACKNOWLEDGMENTS

The author is grateful for financial support received from The Swedish Foundation for Strategic Research.

NOTES

1. For an explanation of how studies of vision differ from expectations, see van der Helm (2009).
2. The CBAM is meant to promote competitive neutrality within and outside the EU, to secure reduction of emissions by limiting the risk of emission transfer from the EU to other parts of the world (SCPC 2023a).
3. In direct reduction, hydrogen is used to reduce hematite into iron sponge. The final product is iron and water vapor: Fe2O3 + 3H2 –> 2Fe + 3H2O, FeO + H2 –> Fe + H2O.
4. Formerly Svenskt Stål Aktiebolag (Public Ltd).
5. Luossavaara-Kiirunavaara Aktiebolag (Public Ltd).

BIBLIOGRAPHY

AFRY Management Consulting. 2022. Fuelling the transition. In *S2E5 Green Steel—Shaping a better future with Eva Blixt, Senior Advisor at Jernkontoret*, edited by Vice President of AFRY Management Consulting Hosted by Matt Brown.

Alriksson, Stina. 2013. "Environmental preferences among steel stakeholders." Linnéuniversitetet, Institutionen för biologi och miljö (BOM).

Andersson, Jenny. 2018. "Arctic Futures—Expectations, interests, claims, and the making of Arctic territory." In *Uncertain Futures: Imaginaries, Narratives, and Calculation in the Economy*, edited by Jens Beckert and Richard Bronk, 83–102. Oxford: Oxford University Press.

Beckert, Jens. 2013. “Imagined futures: Fictional expectations in the economy.” *Theory and Society* 42 (3):219–240. doi: 10.1007/s11186-013-9191-2.

Beckert, Jens, and Richard Bronk. 2018. “An introduction to uncertain futures.” In *Uncertain Futures: Imaginaries, Narratives, and Calculation in the Economy*, edited by Jens Beckert and Richard Bronk, 1–36. Oxford: Oxford University Press.

Bell, Wendell, and A. James Mau. 1971. *The Sociology of the Future: Theory, Cases, and Annotated Bibliography*. New York: Russell Sage Foundation.

Berglund, Henrik, and Dimo Dimov. 2023. “Visions of futures and futures of visions: Entrepreneurs, artifacts, and worlds.” *Journal of Business Venturing Insights* 20:e00411. doi: 10.1016/j.jbvi.2023.e00411.

Berlin, Daniel, Andreas Feldmann, and Cali Nuur. 2022. “Supply network collaborations in a circular economy: A case study of Swedish steel recycling.” *Resources, Conservation and Recycling* 179:106112. doi: 10.1016/j.resconrec.2021.106112.

Borup, Mads, Nik Brown, Kornelia Konrad, and Harro Van Lente. 2006. “The sociology of expectations in science and technology.” *Technology Analysis & Strategic Management* 18 (3–4):285–298. doi: 10.1080/09537320600777002.

Brännberg Fogelström, Julia. 2020. “Experimental study of the temperature profile in an iron ore pellet during reduction using hydrogen gas.”

Brodén Gyberg, Veronica, and Eva Lövbrand. 2022. “Catalyzing industrial decarbonization: the promissory legitimacy of fossil-free Sweden.” *Oxford Open Climate Change* 2 (1). doi: 10.1093/oxfclm/kgac004.

Brown, Nik, Brian Rappert, Andrew Webster, and Stewart Russell. 2004. Contested futures: A sociology of prospective techno-science.

Brunner, Hans-Peter. 2008 “Tipping Points: Managing Complex Systems for Economic Development for Economic Development Success.” Evolutionary and institutional economics review 5 (1): 163–187. doi: 10.14441/eier.5.163.

De La Garza, Alejandro. 2022. “The world’s steel comes at a steep climate cost. A Swedish company is Trying to Change That.” *Time*.

European Commission. 2023. “Carbon Border Adjustment Mechanism.” accessed 2023-08-08. https://taxation-customs.ec.europa.eu/carbon-border-adjustment-mechanism_en.

Excell, Jon. 2023. “Inside Ovako’s hydrogen-fuelled steel plant “ *The Engineer*, July 4.

Fossil Free Sweden. 2023. “Stålindustrin.” accessed 2023-09-10. https://fossilfrittsverige.se/roadmap/stalindustrin/.

Government Offices of Sweden. 2021. “Sweden’s climate policy framework.” https://www.government.se/articles/2021/03/swedens-climate-policy-framework/.

H2 green steel. 2023. “H2 Green Steel in 1.5 billion Euro agreement with ZF.” accessed 2023-08-09. https://www.h2greensteel.com/latestnews/h2-green-steel-in-15-billion-euro-agreement-with-zf.

Hazard, Troy. 2011. *Future-Proofing Your Business Real Life Strategies to Prepare Your Business for Tomorrow, Today*. 1st ed. Hoboken, NJ: Wiley.

IRENA, International Renewable Energy Agency. 2020. Green Hydrogen cost Reduction, scaling up to meet the 1,5°C climate goal.

Jakobsson, Max. 2009. "Från industrier till upplevelser: en studie av symbolisk och materiell omvandling i Bergslagen." Örebro universitet, Akademin för humaniora, utbildning och samhällsvetenskap.

Jansson, Mans, and Goran Ryden. 2023. "Improving Swedish steelmaking: Circulation and localized knowledge-making in early modernity." *Technology and Culture* 64 (2):515–542. doi: 10.1353/tech.2023.0060.

Jernkontoret. 2015. En vision på väg att uppfyllas (Translation: A vision on the way to be realized).

Jernkontoret. 2017. Vision 2050, stål formar en bättre framtid.

Jernkontoret. 2018. Klimatfärdplan—För en fossilfri och konkurrenskraftig stålindustri i Sverige.

Jernkontoret. 2023a. "Beyond Stål." accessed 2023-09-17. https://www.youtube.com/watch?v=yo-gdcCUYVc&list=PLVtBQVnM3DaRbLfQctsaGWkTG6-M9tQwL&index=11.

Jernkontoret. 2023b. "Beyond Stål." accessed 2023-09-10. https://www.jernkontoret.se/sv/vision-2050/stalkreatorer/beyond-stal/.

Johansson, Maria. 2014. "Improved energy efficiency and fuel substitution in the iron and steel industry."**[AQ3]**

Jonsson, Sverker. 2009. *Skuta på stormigt hav: SSAB under tre decennier*. 1. uppl. ed, *Jernkontorets bergshistoriska skriftserie, 45*. Stockholm: Jernkontoret.

Karlsson, Birgit. 2016. "Cartels and norms in the Swedish Steel Industry 1923–1953." *Business History* 58 (7):1077–1094. doi: 10.1080/00076791.2016.1156086.

Komporozos-Athanasiou, Aris, and Marianna Fotaki. 2015. "A Theory of imagination for organization studies using the work of Cornelius Castoriadis." *Organization Studies* 36 (3):321–342. doi: 10.1177/0170840614559258.

Konrad, Kornelia. 2006. "The social dynamics of expectations: The interaction of collective and actor-specific expectations on electronic commerce and interactive television." *Technology Analysis & Strategic Management* 18 (3–4):429–444. doi: 10.1080/09537320600777192.

Kushnir, Duncan, Teis Hansen, Valentin Vogl, and Max Åhman. 2020. "Adopting hydrogen direct reduction for the Swedish steel industry: A technological innovation system (TIS) study." *Journal of Cleaner Production* 242:118185. doi. 10.1016/j.jclepro.2019.118185.

Lenton, Timothy M., Johan Rockström, Owen Gaffney, Stefan Rahmstorf, Katherine Richardson, Will Steffen, and Hans Joachim Schellnhuber. 2019. "Climate tipping points—too risky to bet against." *Nature (London)* 575 (7784):592–595. doi: 10.1038/d41586-019-03595-0.

LKAB. 2023a. "LKAB and the climate." accessed 2023-09-10. https://lkab.com/en/what-we-do/our-transformation/lkab-and-the-climate/

LKAB. 2023b. "Our greatest challenge, your greatest opportunity." accessed 2023-08-20. https://lkab.com/en/.

Luiten, Esther, Harro van Lente, and Kornelis Blok. 2006. "Slow technologies and government intervention: Energy efficiency in industrial process technologies." *Technovation* 26 (9):1029–1044. doi: 10.1016/j.technovation.2005.10.004.

McKay, David I. Armstrong, Arie Staal, Jesse F. Abrams, Ricarda Winkelmann, Boris Sakschewski, Sina Loriani, Ingo Fetzer, Sarah E. Cornell, Johan Rockstrom, and Timothy M. Lenton. 2022. "Exceeding 1.5°C global warming could trigger multiple climate tipping points." *Science (American Association for the Advancement of Science)* 377 (6611):1171-eabn7950. doi: 10.1126/science.abn7950.

Mische, Ann. 2009. "Projects and possibilities: Researching futures in action." *Sociological Forum (Randolph, NJ)* 24 (3):694–704. doi: 10.1111/j.1573-7861.2009.01127.x.

Nwachukwu, Chinedu Maureen, Elias Olofsson, Robert Lundmark, and Elisabeth Wetterlund. 2022. "Evaluating fuel switching options in the Swedish iron and steel industry under increased competition for forest biomass." *Applied Energy* 324:119878. doi: 10.1016/j.apenergy.2022.119878.

Öhman, Amanda, Emrah Karakaya, and Frauke Urban. 2022. "Enabling the transition to a fossil-free steel sector: The conditions for technology transfer for hydrogen-based steelmaking in Europe." *Energy Research & Social Science* 84:102384. doi: 10.1016/j.erss.2021.102384.

Oomen, J. J., J. G. Hoffman, M. A. Hajer, Futures Urban, and Geowetenschappen Faculteit. 2022. "Techniques of futuring: On how imagined futures become socially performative." *European Journal of Social Theory* 25 (2):252–270. doi: 10.1177/1368431020988826.

O'Riordan, Timothy, and Tim Lenton. 2013. *Addressing Tipping Points for a Precarious Future, British Academy Original Paperback*. Oxford: Oxford University Press.

O'Riordan, Tim, Tim Lenton, and Ian Christie. 2013. "Metaphors and systemic change." In *Addressing Tipping Points for a Precarious Future*, edited by Timothy O'Riordan and Tim Lenton, 3–20. Oxford: Oxford University Press.

Pei, Martin, Markus Petäjäniemi, Andreas Regnell, and Olle Wijk. 2020. "Toward a fossil free future with hybrit: Development of iron and steelmaking technology in Sweden and Finland." *Metals (Basel)* 10 (7):1–11. doi: 10.3390/met10070972.

Pettit, Krista L., Julia Balogun, and Mandy Bennett. 2023. "Transforming visions into actions: Strategic change as a future-making process." *Organization Studies*:17084062311718. doi: 10.1177/01708406231171889.

Porter, A. J., Timothy Kuhn, and Brigitte Nerlich. 2018. "Organizing authority in the climate change debate: IPCC controversies and the management of dialectical tensions." *Organization Studies* 39 (7):873–898. doi: 10.1177/0170840617707999.

Rockström, Johan. 2023. Keynote at Swedish Steel Prize.

Ruef, Annette, and Jochen Markard. 2010. "What happens after a hype? How changing expectations affected innovation activities in the case of stationary fuel cells." *Technology Analysis & Strategic Management* 22 (3):317–338. doi: 10.1080/09537321003647354.

Sandberg, Hans, Rune Lagneborg, Birgitta Lindblad, Helén Axelsson, and Lars Bentell. 2001. "CO_2 emissions of the Swedish steel industry: CO_2 emissions of the Swedish steel industry." *Scandinavian Journal of Metallurgy* 30 (6):420–425. doi: 10.1034/j.1600-0692.2001.300611.x.

SCPC, the Swedish Climate Policy Council. 2023a. The annual report of the Swedish Climate Policy Council.

SCPC, the Swedish Climate Policy Council. 2023b. "By 2045 Sweden shall have no net greenhouse gas emissions." accessed 2023-08-20. https://www.klimatpolitiskaradet.se/en/.

SCPC, the Swedish Climate Policy Council. 2023c. "Report 2023." accessed 2023-08-20. https://www.klimatpolitiskaradet.se/en/report-2023/.

SSAB. 2022a. Annual report—Leading the green transition of the steel industry.

SSAB. 2022b. "Reinventing the future of steel." https://www.ssab.com/en.

SSAB. 2023. "SSAB and BE Group announce partnership for fossil-free steel." SSAB Newsroom, accessed 2023-08-20. https://www.ssab.com/en/news/2023/05/ssab-and-be-group-announce-partnership-for-fossilfree-steel.

Stern, Nicholas. 2023. "Endorsement." accessed 2023-08-08. https://www.systemiq.earth/wp-content/uploads/2023/01/The-Breakthrough-Effect.pdf.

Stirling, Andy. 2015. "Emancipating transformations—From controlling the transition to culturing plural radical progress." In *The Politics of Green Transformations*, edited by Ian Scoones, Melissa Leach and Peter Newell, 54–67. New York: Routledge, Earthscan.

Swedish Energy Agency. 2023. "The industrial leap." accessed 2023-08-20. https://www.energimyndigheten.se/en/innovations-r--d/energyintensive-industry/the-industrial-leap/.

Systemiq. 2023. "The breakthrough effect: How to trigger a cascade of tipping points to accelerate the net zero transition.." In *Systems Change Lab with the Support of Bezos Earth Fund*: Developed by Systemiq in partnership with the University of Exeter and Simon Sharpe.

Tang, Ou, Jakob Rehme, Pontus Cerin, and Donald Huisingh. 2021. "Hydrogen production in the Swedish power sector: Considering operational volatilities and long-term uncertainties." *Energy Policy* 148:111990. doi: 10.1016/j.enpol.2020.111990.

Tutton, Richard. 2017. "Wicked futures: Meaning, matter and the sociology of the future." *The Sociological Review (Keele)* 65 (3):478–492. doi: 10.1111/1467-954X.12443.

van der Helm, Ruud. 2009. "The vision phenomenon: Towards a theoretical underpinning of visions of the future and the process of envisioning." *Futures: The Journal of Policy, Planning and Futures Studies* 41 (2):96–104. doi: 10.1016/j.futures.2008.07.036.

van Lente, Harro. 2012. "Navigating foresight in a sea of expectations: lessons from the sociology of expectations." *Technology Analysis & Strategic Management* 24 (8):769–782. doi: 10.1080/09537325.2012.715478.

Vogl, Valentin, Johan Rootzén, and Oscar Svensson. 2019. "A just transition towards a coal-free steel industry: perspectives from labour."

We mean business. 2022. Making Net-Zero S teel Possible, An industry-backed, 1.5°C-aligned transition strategy.

We mean business coalition. 2023. "The latest from We mean business coalition." accessed 2023-08-20. https://www.wemeanbusinesscoalition.org/blog/net-zero-transition-the-latest-signals-of-change-july-21-2023/.

Wenzel, Matthias, Hannes Krämer, Jochen Koch, and Andreas Reckwitz. 2020. "Future and organization studies: On the rediscovery of a problematic temporal category in organizations." *Organization Studies* 41 (10):1441–1455. doi: 10.1177/0170840620912977.

Whiteman, Gail. 2023. "Endorsement." accessed 2023-08-08. https://www.systemiq.earth/wp-content/uploads/2023/01/The-Breakthrough-Effect.pdf.

Chapter Three

Best Control Versus Best Available Technology

A Precarious Link Towards Safeguarding Environmental Justice

Frederick D. Gordon

Environmental justice counts. However, the way it is presented may provide insight into best-practice solutions. Therefore, this chapter is divided in the following manner. 1) A domestic policy overview, 2) international policy framework, 3) role of democratization, 4) state-society relations, and 5) Best Available versus Best Control Technology overview. This chapter argues that applying domestic policy constructs and adaptations is only as effective as unimpeded policy implementation options. It supports the claim that these developments are contingent on the role of democracy, rule of law, and transparent government practices. Only then will the most vulnerable groups have the best chance of obtaining reform. Recent environmental justice literature has been directed toward state-centered solutions, as evidenced by multilateral support from international donor agencies such as the World Bank and International Monetary Fund. While there is a lot of plausibility to this approach, the reality is that much of the developing world remains mired in poverty, with many per capita households still hovering around two dollars a day. Closer examination reveals that people lack access to freshwater, succumb to numerous diseases, experience high levels of infant mortality, and lack of opportunity to improve individual and group welfare.

DEFINING ENVIRONMENTAL JUSTICE

Environmental justice has numerous definitions. An international framework suggests that it reflects impoverished living conditions, lacking in essentials like freshwater, food, and medicines. A domestic interpretation is in general agreement but there is some variation. For instance, domestic environmental justice concerns have been linked to a more strident term called environmental racism. This phrase can be defined as "any policy, practice of directive that differentially affects or disadvantages (whether intended or unintended) individuals, groups or communities based on race or color" and is used to explain the differential treatment of populations on environmental issues.[1] Yet race is a difficult proposition to tackle because it is hard to prove. Recent studies also suggest that it may be more advantageous to focus more on solutions rather than name classifications.[2]

Regardless, people of color and other low-income groups have been disproportionately exposed to environmental harms and toxins. The legal pathway for reform is quite complicated. One would need to prove a Title VI violation, which requires the level of intent. Many of the domestic environmental justice claims that have arisen reflect the likelihood of people of color or other low-income groups being negatively impacted by industrial pollutants. While this may reflect correlation, it does not meet the standard of causation.[3] International law is the basis for solving international environmental justice claims. However, there are other challenges since international law or "soft law" is hard to enforce and may of the actual impact sites are located non-democratic countries which lack basic human rights safeguards and provisions. It appears that the domestic and international constructs of environmental justice are far apart. However, this is a misnomer. Their premises are very similar in that no one should be denied access to live in a healthy environment. The main distinction is that domestic environmental justice concerns occur in countries which adhere to the rule of law and promote basic rights, either as mature or consolidating democracies. At the international level, many of the countries which house vulnerable populations do not exhibit these democratic checks and balances.

DOMESTIC POLICY

In this section, I review the origins of the environmental justice movement, role of regulations, national environmental policy developments, and the rule of law. It could be said that an injustice anywhere is an injustice everywhere. This aphorism applies to the concept of environmental justice.

However, questions remain as to whether it should be viewed in an ominous way. For some the answer is yes, that low-income people have undoubtedly been exposed to harmful toxins. This has resulted in poorer quality of life including asthma and other breathing-related conditions. In some ways, this can be viewed as a negative externality of modern development. Society has industrialized and brought it with factories and other manufacturing entities into local neighborhoods. In turn, they have supplanted natural pristine environments with factories that pollute the atmosphere. This has led to a series of calls of injustices which infers racial-based or class-based animus from distressed communities. The industry sees it differently in that they never had intent to harm or cause societal dismay. In fact, they often believe that their industries only play a nominal role in impacting community life. The issue becomes trifold because governmental leaders have often been charged with doing too little too late.

Environmental justice is a domestic rights approach and leans on the 1964 Civil Rights Act for legal support. And in some ways, it has been effective. Today the Federal government supported by executive legislation requires all federal agencies to address environmental justice concerns. Critics chime that the environmental justice Executive Order 12898 signed under Bill Clinton has not been effective enough in mitigating environmental justice problems. President Clinton had reframed the debate by passing an executive order which mandated each federal agency establish an office to address environmental justice claims. This executive decree has received mixed results because, while it recognizes the need to address environmental harm, its efforts have been disjointed which in turn have produced uneven results in terms of effective cleanup.[4]

While this debate is far from settled, it is clear that the federal government has made more than modest efforts to address this concern. However, this has not always been the case in terms of both environmental policy and justice. Environmental policy in the United States can be classified as bifurcated and contentious. In the United States, there is no environmental amendment, no environmental cabinet position and legislative reform did not truly begin until the 1970s. It was only with the establishment of the National Environmental Policy Act in 1969 that Congress took a leadership role in protecting the nation's water and air quality. This was evidenced with the establishment of the Environmental Protection Agency in 1970 and passage of the Clean Air Act (1970) and Federal Water Pollution Control Act in 1972. However, the advent of the environmental justice movement still lagged behind another decade.

The 1970s became known as an environmental decade, featuring numerous legislative reforms directed at addressing industrial responsibility, cleanup, species protection among other protective measures. These efforts were championed by environmentalists, pro-regulation activists, but were met with steep resistance from industry and anti-regulation factions. Yet the dawn of environmental justice surfaced in the early 1980s at a time when environmental reforms were being rolled back.

The first national environmental justice protest occurred in 1982 over the planned disposal of polychlorinated biphenyl (PCB) wastes in Warren County, North Carolina. Widespread allegations of inequality in the distribution of environmental risks emerged on the national scene in 1987 with the release of the study by the United Church of Christ. The study "Toxic Wastes and Race in the United States," reviewed the demographic characteristics around commercial hazardous waste facilities and found that race was the most significant factor in predicting the likelihood of living near such a facility—communities with the greatest number of commercial hazardous waste facilities had the highest composition of racial and ethnic residents.[5] Yet, new studies stretch further back and suggest that it was linked to Jim Crow policy. Most notably, are recent publications depicting the black refugee plight after the Great Mississippi Flood of 1927. Blacks received minimal relief support compared to their white counterparts. Flood victims were forced to live in deplorable conditions, all the while the alleged was administered by the Red Cross.[6]

In the former case, presidential leadership offered a minimalist approach in addressing environmental justice concerns. It was a different time, and the role of the federal government was limited. In fact, it was perceived that the private sector should be the responsibility to assist flood victims. Flash forward half a century, the role of the federal government was stridently different. In 1994, President Clinton issued Executive Order Number 12898, "Federal Actions to Address Environmental Justice in Minority Populations and Low-Income Populations," and adopted the phrase "environmental justice" to refer to disproportionately high and adverse human health or environmental effects on minority populations and low-income populations. Rather than explicitly defining the phrase, the executive order elaborated on its meaning by requiring each federal agency to develop strategies to achieve environmental justice by, at a minimum, (1) identifying and addressing disproportionately high and adverse human health or environmental effects of agency programs, policies and activities on minority populations and low income populations 2) promoting enforcement of all health and environmental statutes in areas with minority or low-income populations 3) ensuring greater public participation 4) improving research and data collection relating to the health and environment of minority and low income populations and 5) identifying differential patterns of consumption of natural resources among minority and low income populations.[7] It seems that the executive order was a solutions-based decree. Whether that could be achieved links back to legal interpretation.

The 1964 Civil Rights Act was passed as a measure to protect against discriminatory behavior. However, the key was in Title VI which required one to prove intent. This meant that there had to be clear and convincing behavior to prove discrimination. Domestic environmental justice claims were challenged by reaching the level of intent. However, there is a secondary level known as disparate social impact. This means that there is a likelihood for harm to occur but falls short of intent. This issue was addressed by the high court as was the case with the 2001 Supreme Court Decision of Alexander v. Sandoval. It was not even an environmental justice case, but the concept of disparate social impact was on trial. In this case, section 601 of Title VI of the 1964 Civil Rights Act was in question. The section prohibits discrimination based on race, color, or national origin. Respondent Sandoval brought a class action to enjoin Alabama's Department of Public Safety's decision to administer state driver's license examinations only in English, arguing that it violated the DOJ regulation because it had the effect of subjecting non-English speakers to discrimination based on their national origin. Relief was temporary. The District Court enjoined the policy and ordered the Department to accommodate non-English speakers. The Eleventh Circuit affirmed as well. However, the US Supreme Court held that that there is no private right of action to enforce disparate-impact regulations promulgated under Title VI. The Court was tempered to extend a disparate impact claim to fall under Title VI. The main concern was the fear that such an extension could result in a bevy of suits.[8] This also could be viewed as a pro-business decision because an overabundance of disparate impact claims could potentially have an unwanted regulatory impact.

One of the most perplexing things about environmental justice is the availability of readily accessible solutions. In the domestic realm, possible remedying solutions range from pollution abatement to civil rights redress. However, another crucial aspect of management that must be included is societal input. Too often, vulnerable groups feel disenchanted from trusting the political process, leading them to eschew political involvement. Secondly, these groups may be dependent upon local industry for employment; as a result, they are hesitant to comment on pollution problems. Yet, according to Kuehn, societal involvement is exactly what is needed. He identifies four vantage points that encourage group collective action and frames them as 1) distributive justice, 2) procedural justice, 3) corrective justice, and 4) social justice. In essence, he builds on the work of Bullard, not only identifying root cause of environmental justice but also offering cogent policy solutions.[9]

The first approach distributive justice has been defined as "the right to equal treatment that is to the same distribution of goods and opportunities as anyone else." Distributive justice involves the equitable distributions of the burdens resulting from environmentally threatening activities or of the environmental benefits of government and private sector programs. More specifically, distributive justice involves the addressing the disproportionate public health and environmental gaps and risks borne by people of color and lower income groups. It hints at a Rawlsian perspective in that societal burdens must be equally distributed. Yet, this idea also closely parallels a utilitarian approach in seeking the greatest good for the greatest number.[10]

Second, procedural justice has been defined as "the right to treatment as an equal." The emphasis here is not so much on the requirement of an equal distribution of goods or services but rather on an equal role in the decision-making process in how these goods and opportunities are to be distributed. In some ways, this way of thinking extends back to the ancient Greeks. Aristotle referred to this as individuals who have an "equal share in ruling and being ruled." More specifically, this means the desire to create fairness in decision making. According to Bullard, this is known as "procedural equity" or the need for democratic decisions which encompasses inclusiveness, representation, parity and communication.[11]

Third, corrective justice involves fairness in the way punishments for lawbreaking are assigned and damages inflicted on individuals and communities are addressed. Aristotle referred to this aspect of justice as "rectificatory" as "it treats the parties as equals and asks only whether one has done and the other has suffered damage." It attempts to restore the victim to the condition she was in before the unjust activity occurred. Corrective justice involves not only just administration, but also a duty to repair losses for which one is responsible. This is most reflected in Federal Executive Order 12898, which applies corrective justice by directing agencies to develop strategies to promote enforcement of health and environmental statutes in minority and low-income populations and to collect, maintain and analyze information on the race, national origin, and income of populations surrounding facilities or sites that become the subject of a substantial federal enforcement action. EPA's environmental justice definition encompasses corrective justice concerns in calling for "fair treatment with respect to the development and enforcement of environmental laws, regulations and policies.[12]

The fourth and final aspect of environmental justice is "social justice" a far reaching and some say a distant goal. Social justice is "that branch of the virtue of justice that moves us to use our best effort to bring about a more just ordering of society one in which people's needs are more fully met." The social justice demands every class have enough resources and enough power to live as befits human beings, and second, that the privileged classes are accountable to society."[13]

Environmental justice can be viewed as a "merger with environmentalism" integrating environmental concerns into a broader agenda that emphasizes social, racial and economic justice. Bullard referred to this aspect of environmental justice as "social equity" an assessment of the role of sociological factors (race, ethnicity, culture, lifestyles, and political power) in environmental decision-making.[14]

INTERNATIONAL ENVIRONMENTAL JUSTICE

In this section, I offer an explanatory overview of international environmental justice and how it developed as well as the legal parameters to ensure its protection. International and domestic environmental justice are similar in that they both seek to limit environmental degradation as well as support *intergenerational* equity. They differ in terms of rule of law and overall commons protection. In general, domestic environmental justice is supported by the federal government. This features the use and application of Executive Order 12898 and use of federal mandates to ensure that states comply with federal rules and regulations. In comparison the international environmental justice framework is grounded in international environmental law and treaties. One of the most noted examples of international efforts and cooperation has to do with climate change. It's a front-line environmental justice issue because vulnerable people and groups are highly susceptible to climate change. A slight temperature increase could make an already vulnerable situation inhabitable. The problem is further exacerbated because such groups and individuals have limited resources at their helm to rectify the situation. This means that they could end up as refugees or even worse as internally displaced. Further, many of the international impact sites are in locations that are ruled by strong authoritarian regimes and where the rule of law is in question. Disputes at the international level also include allegations that governments and multinational corporations exploit indigenous peoples and the impoverished conditions of developing nations.

Some people say that redress for international environmental justice claims is late to the party. After all it is positioned with the fourth generation of international human rights. Scott posits that other core rights like political and economic were sanctioned much earlier. It was only in the later part of the 1980s that human rights activists clamored for additional reforms and began to promote the rights of indigenous peoples. This inferred that local groups predominantly located in developing regions would receive some level of international support and protection. Yet it was unclear how environmental rights would stand in comparison to the more established political and civil rights. However, international legal gains were made by homing in on protecting indigenous groups and the right to a healthy environment.

According to Scott, human rights can be classified in terms of "generations." She posits the human rights can be broken down decadal and in a linear manner. She states civil and political rights are referred to as *first generation* human rights. This was the first human rights advancement after establishing the Universal Declaration of Human Rights in 1948.[15] Two decades later, economic, social and cultural rights were championed as *second-generation* rights. This was an important finding because it demonstrated that people had a right to basic economic needs and protection. Scott offers unusual insight into the origins of these rights. She identifies political and civil rights as linking back to the eighteenth-century Declaration on the Rights of Man. Yet she sees a divergent discourse in terms of economic, social and cultural rights. It seems like it's closely linked to historical determinism in that these rights are conditioned based on location. Scott posits that such rights have strong ties to Cold War politics in that former Soviet Union countries much closer to such rights than their western counterparts.

Human rights law is multi-tiered and has made marked gain in response to war, famine, societal displacement and genocide. Yet, international legal maturity was distinguished by recognizing the rights of both individuals and groups. This was clearly evidence in the 1980's. *Third generation rights* link to group protection and affirmation and most notably are referred to as right to self-determination and the right to development. The right to development was declared by the General Assembly in its 1986 Declaration on the Right to Development and recognized in the African Charter on Human and Peoples' Rights. Article 24 of the African Charter on Human and Peoples' Rights refers to the right of peoples "to a general satisfactory environment" was referred to in the Additional Protocol to the American Convention on Human Rights in the Area of Economic, Social and Cultural Rights. The right to development and self-determination are powerful social constructs. They implicitly support economic advancement and with self-determination potential statehood.[16] A prevailing theory is that peoples' general welfare would improve if they were not merely subjects of distant, detached or repressive rule of law.

It was not until the close of the Cold War, that nation state protocols began to shift attention to the most vulnerable groups and the environment. It may have been a paradigm shift from the harsh neo-realist framework which looked at a bipolar global arena in austere terms. High security issues prevailed and hence concepts like indigenous groups and the environment could not compete with power concepts like Mutual Assured Destruction or deterrence theory. However, as the Cold War waned, greater attention was shifted to recognize other less salient issues, which in turn paved the way for the *fourth-generation rights*.[17]

This directly refers to indigenous peoples who were not expressly included in the text of any United Nations human rights instrument before the 1989 Convention on the Rights of the Child; article 30 of that convention provides that a child who is indigenous shall not be denied the right in the community with other members of his or her group to enjoy his or her own culture to profess and practice his or her own religion or to use his or her own language. There was some precedent. This International Labor Organization Convention 107 issued Article 2 that afforded governments the "primary responsibility for developing coordinated and systematic action for the protection of the populations concerned and their progressive integration into the life of their respective communities."[18] It seems that these developments were early incidents for the group protection.

Ultimately, it was not until 2007, when the UN Declaration in the Rights of Indigenous Peoples was adopted "as a standard of achievement to be pursued. Article 1 recognizes that indigenous peoples have the right to full enjoyment in a collective or as individuals, of all human rights in international human rights law." Articles 3 and 4 recognized the right of indigenous peoples to self-determination and to autonomy or self-government in matters relating to their internal and local affairs.[19] The link to environmental justice is clear because indigenous peoples are some of the most vulnerable groups and subject to high rates of disease, food deprivation, climate change impact amongst other encroachments to their way of life.

Yet, despite some significant international rights gains, there is a plethora of critics who see environmental justice claims as unfulfilled. Yokata's study reveals that *intergenerational* equity and international justice must be emphasized more. And even if one is not looking to the long term, policymakers must act now to protect citizenry from harmful effects attributed to global warming and the depletion of the ozone layer. In essence, ecocriticism points are raised by pointedly stating that plain and simple it is unjust to watch wildlife drown in oil spills, or to allow entire species to disappear at the hands of industrialization. While Yokata does not openly decry environmental justice violations, domestic policy considerations that policy recommendations must extend beyond merely identifying an issue. Rather, it must offer just recourse and compensation to people victimized by environmental degradation and develop a substantive legal framework.[20]

Yokata, believes the United Nations Charter is partially at fault because it does not clearly define justice. It speaks of the determination "to establish conditions under which justice and respect for the obligations arising from treaties and other sources of international law can be maintained." It also stipulates that one purpose of the United Nations is "to bring about by peaceful means, and in conformity with the principles of justice and international law, adjustment or settlement of international disputes or situations which might lead to a breach of the peace." Furthermore, the charter provides that all members shall settle their international disputes by peaceful means in such a manner that international peace and security and justice are not endangered. However, it is not clear whether justice means anything more than justice between states. This is an important finding because it could mean that effective policy solutions remain within the state. One might think the state would always seek ecumenical solutions. However, that approach is ajar with political realism which posits that states seek to maximize their self-interest. This approach could minimize the rights of indigenous peoples and their environment.[21]

PARADIGM SHIFT

Some political science scholars have advanced the concept of a paradigm shift as a way to safeguard international environmental justice concerns. Chasek, Downie, and Brown offer important insight in Global Environmental Politics. They present a continuum of international treaties which feature lead states, blocking states and supporting states. By this it shows a connection in how and why some treaties are fully operationalized while others are not. International cooperation is often predicated on who is the lead state and who are the blocking or veto states. In terms international environmental accord, efficacy, treaties warrant multiple layers of cooperation, and this takes time.[22] How and when that occurs is a more complex phenomenon and perhaps it can be best viewed through a paradigm shift. According to Rodman, every political paradigm assumes that the purpose of political activity is to promote, either directly or indirectly, the good life. The crucial differences stem from differing conceptions of the good life and how explicit they are. Classical (ancient and medieval) political theorists were very explicit. For example (e.g., Aristotle's analysis of socioeconomic roots of civil instability) was to identify the conditions that would foster the good or happy life or man, defined as activity in accordance with virtue or excellence.[23] This means that there must be a level of agreement on common goals and overall collective action.

The transition from the classical to the modern paradigm is fraught with subtle inconsistencies. On one level, the politics of virtue becomes discredited and supplanted by a more worldly and productive politics of prosperity and wealth. On a subtler level human virtue, classically construed not only as transcendence of nature but also as a repression of the nature develops in an increasingly technological age into full-fledged vision of human realization via the domination and transformation of external nature. As talk of virtue diminishes, the tacit image of the virtuous man shifts from the sage and the monk to the entrepreneur and the technocrat. This passage is quite revealing for a number of reasons. First it offers a constructionist approach to how individuals or leaders make decisions and further offers insight into policy boundaries and how self-interest could trump the public good.[24]

DEMOCRATIZATION

In this section, I explore the nexus between democratization and international environmental justice. There is a long ledger of human rights developments to support and protect individuals and groups. However, another critical feature which must be considered is the role of democratization. In all fairness, democratization is a complex concept and is even more obfuscated since many of the most glaring environmental justice issues remain inside the borders of authoritarian and dictatorial regimes. Concepts like paradox of plenty and clientelism are sad reminders of how peoples' basic rights can be abrogated. However, one should not think the role of democratization is a guarantee to address all environmental justice claims either first, nearly half of the global community is not democratic and even on the democratic side, there are different levels of democracy. There are mature democracies which have established rules of law and yield a transparent change of power system. There are also democratically consolidating countries. This means that while they are on the path to democratic reform, it is not complete. Hence, one might expect less attention to be spent on protecting indigenous groups and the environment. Mature democracies may aid in the paradigm shift because countries that have a full application of civil rights and liberties protection may be best suited to address environmental justice claims. Diamond states a mature liberal democracy must feature the following four concepts: 1) civil rights and liberties, 2) free press, 3) open elections, and 4) a multi-party system.[25] These are all valuable points and could be seen supportive ways for enhancing environmental justice causes. It follows that effective international environmental accords like climate change must feature lead democratic countries and set standards for international compliance.

STATE SOCIETY RELATIONS

In this section, I advance the claims of the state and suggest that a paradigm shift must extend beyond the transition to democracy. It must also include a thorough examination of state-society relations. Migdal offers critical insight in understanding the state's transformation by hinting that there must be a recursive relationship between state and society. This means that the state must listen to its people and that society must listen to the state. This is a complex phenomenon and an entirely different approach in addressing international environmental justice claims. However, it could be a game-changer if this relationship remains present.[26]According to Migdal, this has not always been the case. He states, "the modern state has posed itself as the ultimate authority, but the engagement of social groups within the state, have tempered those broad claims." Migdal qualifies this assessment and notes that states may be badly fractured and weakened through the particular nature of their encounters with other social forces. He believes state behavior has been modified through globalization and other twenty-first-century encounters."[27] This could mean identifying other international events and possibly creating an aperture for the recognition of fourth generation of rights.

In all fairness, the connection between the state and its population is complex. This may have to do with power politics. Migdal states, "the modern state has demanded much from its subjects." If, in premodern empires, rulers might have aimed for little more than revenue from many peasants, modern officials have devised legal codes, specifications of what individuals can and cannot do." Yet states have been unable to transform societies sufficiently so as to solve the paradox of being simultaneously apart from society and a part of society."[28] More than that, the engagement of the state with society has led to contending coalitions that have both cut across and blurred the lines between them. This may be the ultimate paradigm shift that is required to safeguard environmental justice concerns because the state cannot operate in an isolated state manner. Any legitimate state needs to have a strong recursive relationship with its population, both rural and urban. Otherwise, conditions could be ripe for internecine warfare or regime change. Both of which are deadly impediments to addressing environmental justice causes.

BEST CONTROL VERSUS BEST AVAILABLE TECHNOLOGY

In this section, I advance the argument that technological approaches must also be considered. This can be advanced through democratization and effective recursive state-society relations. This is promising but to stop there would be an incomplete analysis. Environmental problems can be ameliorated by applying effective technology. It can be evidenced in terms of having access to running water, medicine, and a sufficient food supply. Of course, such ideas are contingent upon the type of government as well as clear implementation strategies. This chapter has supported numerous approaches in addressing environmental justice claims, but technology also counts which includes best available and best control strategies. This is a challenge because these concepts are a function of the Federal Water Pollution Control Act and are used in preventing unhealthy discharge into US rivers and tributaries. Hence, a seminal question arises which is whether a domestic construction can be applied at the international level?

Kubasek yields insight into what is exactly at stake (Kubasek & Silverman, 2002). She describes the importance of such technology. Best control technologies have been devised to impact discharge into surface water and protect and against pollutants from sources such as factories, pulp, and paper mills, food processing plants and wastewater treatment plants. The Environmental Protection Agency also notes that over 200,000 sources are regulated by these permits. These permits are administered through the National Pollution Discharge Elimination System (NPDES) process and dischargers must follow specifications of their NPDES permits. Effluent limitations are dependent on the type of pollutant, type of discharger, and characteristics of the local watershed. The standards are *conventional, non-conventional and toxic.* Requirements vary depending on whether the discharge is a new or existing source. Furthermore, municipal wastewater treatment plants must meet a different level of standards.[29]

The least stringent level of control is imposed on existing dischargers of conventional pollutants requiring only that best conventional pollutant control technology (BCT) standards are met. Only five conventional pollutants are specified, biochemical oxygen demand (BOD), total suspended solids, pH and fecal coliform were established in the 1977 amendments and the EPA subsequently added oil and grease.[30]

In establishing BCT standards, the EPA is required to consider the relationship between the pollution control costs and derived benefits; those costs must then be compared to the cost of compliance in wastewater treatment plants. In addition, the EPA is required to consider the age of equipment and facilities involved, the process employed by the engineering aspects of the application of various types of control techniques, process changes, non-water quality impact (including energy requirements), and such other factors as the Administrator (of the EPA) deems appropriate.[31]

Toxic and non-conventional pollutants must be controlled using best available technologically economically achievable (BAT), a more rigorous standard than BCT. Non-conventional pollutants are anything not included in these other two categories that still may pose a threat to water quality (such as thermal pollution). A toxic pollutant list was developed by the EPA. The list includes individual chemicals and chemical groups. In establishing effluent limitations, the EPA can establish specific chemicals used by that industry. For example, the new source performance standards for steam electric power generating facilities identifies a priority pollutant list of 126 specific chemicals for which none should be detectable.[32]

BCT, which has the most stringent standards reveals a complex and dynamic cost benefit analysis configuration. It takes into consideration the concept of reasonableness in attaining a reduction in effluence benefits. BAT standards only require that standard determination shall take into account the cost of achieving such effluent reduction without identifying specific benefits. This reduced requirement for cost assessment has resulted in the EPA establishing BAT standards primarily based on health criteria. The 1977 amendments to Clean Water Act are straightforward in that "the toxicity of the pollutant, its persistence, degradability, the usual or potential presence of the affected organisms in any waters, the importance of the affected organisms and the nature and extent of the effect of the toxic pollutant on such organisms, and the extent to which effective control is being or may be achieved under other regulatory authority."[33]

New sources have more rigorous effluent limits than existing sources. This is based on the belief that it is cheaper to minimize effluent pollutants if environmental controls are considered during plant design than it is to retrofit existing facilities. New sources include major modifications to existing facilities, as well as new construction. Effluent limits for new sources are based on consideration of the entire process rather than the BAT and BCT approach, which requires pollution controls to be imposed following waste generation (commonly known as end of pipe treatment). As specified the FWPCA standards must consider control of the discharge of pollutants which reflects the greatest degree of effluent reduction which the Administrator determines to be achievable through application of the best available demonstrated control technology, processes, operating methods and other alternative including where practicable, a standard permitting no discharge of pollutants.[34]

The question—how could such a domestic policy concept apply to such a broad international concept? While these control mechanisms seemingly have been labeled as domestic constructs, there is important application at the international level for both BAT and BCT. Both approaches have been used to address effluent discharge from rivers and waterways. It many ways, these approaches have been designed to address point source pollution and they are not well suited to address non-point source pollution concerns. However, they also offer a new way of thinking. BAT is a current way of thinking in that one works with what they have. It's kind of like treating the symptoms. BCT goes further and identifies the root cause, which is what is so essential in addressing international environmental justice claims. BCT technology posits that there are no societal barriers which should inhibit the best pollution abatement practices. It also leaves little to imagine because it is a standard designed to combat toxic discharge. This means that the risk must be handled comprehensively.

Yet could this be applied to basic human rights and further help alleviate environmental justice claims? Ultimately, the best technological solutions must be precise and have a direct impact. In the case of climate change, which is another front-line environmental justice concern, the concept of clear development mechanisms has been devised, which allows developed countries to help developing countries produce better equipment to reduce their global footprint of climate change. In fact, the Clean Development Mechanism (CDM), defined in Article 12 of the Kyoto Protocol allows a country with an emission-reduction of emission-limitation commitment, to implement an emission reduction project in developing countries. Such projects can earn saleable certified emission reduction credits, each equivalent to one ton of carbon dioxide.

The same approach could be applied by creating technology to limit developing countries' pollution footprint. Of course, much of this analysis is heavily contingent on the type of government involved, the role of democratization as well understanding the dynamics of healthy state-society relations. Yet too much is at stake not to consider such an approach.

CONCLUSION

This chapter presented a novel argument that more needs to be done to address international environmental justice concerns. It offered a comparison between domestic environmental justice concerns. Both levels seek redress from environmental "bads." How that is achieved is another matter. Domestic Environmental Justice claims rely on the rule of law, the role of the federal government, and then emphasize a litany of societal measures to seek redress. This can be a daunting concern since it is very difficult to prove a case because intent is required.

The international level is equally challenging. It is grounded in international law, and while it fits with the fourth generation of rights, true efficacy must factor in the role of democratization and state society relations. In short, this means that there must be a legal rational rule of law process as well as a healthy level of communication between the state and its inhabitants. If either of these approaches is missing, then it could prove problematic to address international environmental justice claims. The forecast is questionable because many of the most glaring environmental justice issues are located in politically unstable areas. Therefore, as the chapter suggests, a paradigm shift must occur which means that the state must rethink its roles, rights and obligations to its citizenry. Secondly, more attention must be directed toward implantation. There needs to be effective leaders locally and nationally who desire to promote change. Technology alone will not suffice, but it could be a good starting point. Measures have been passed designed to bridge the developing-developed state chasm. Those efforts must continue, and increased attention must be directed toward the state in deriving and exhibiting strong relations with its citizenry. Treaties and accords count but so do the most vulnerable and impoverished. The state cannot afford to allow them to remain silent. There needs to be a bridge for current leaders to make this tacit connection. Only then will concepts intergenerational equity have meaning. It's a precarious proposition, but then again there really are no other options.

NOTES

1. Robert R. Kuehn. "A Taxonomy of Environmental Justice," The University of Alabama School of Law, *Environmental Law Reporter*, vol. 30 (2000), p. 10681.
2. Ibid.
3. Nancy K. Kubasek and Gary S. Silverman, *Environmental Law*, Fourth Edition (Upper Saddle River: Pearson, 2002), pp. 33–34.
4. David Kosinsky, editor, *Failed Promises: Evaluating the Federal Government's Response to Environmental Justice (American and Comparative Environmental Policy)* (Cambridge: MIT Press, 2015), pp. 1–15.
5. Kuehn, "A Taxonomy of Environmental Justice," p. 10682.
6. Frederick D. Gordon. *The Damming of the Presidency* (Lanham: Rowman and Littlefield, 2020), pp. 153–154.
7. Kuehn, "A Taxonomy of Environmental Justice," p. 10682.
8. Alexander v. Sandoval, 532 U.S. 275 (2001) in Alexander v. Sandoval: 532 U.S. 275 (2001): Justia US Supreme Court Center, last date accessed, 22-3-2021.
9. Robert D. Bullard, *Unequal Environmental Protection: Incorporating Environmental Justice in Decision Making*, Adam M. Finkel and Dominic Goldings (eds.) (1994).
10. Robert R. Kuehn, "A Taxonomy of Environmental Justice," pp. 10682–10684.
11. Ibid.
12. Ibid.
13. Ibid.
14. Ibid.
15. Shirley Scott, *International Law in World Politics: An Introduction*, Second Edition (Boulder: Rienner, 2010), pp. 230–231.
16. Shirley Scott, *International Law in World Politics: An Introduction*, pp. 230–231.
17. Ibid.
18. Ibid.
19. Ibid.
20. Yozo Yokota. "International Justice and the Global Environment," *Journal of International Affairs*, 52:2, 1999, pp. 583–589.
21. Yozo Yokota, "International Justice and the Global Environment," pp. 583–589.

22. Pamela S. Chasek, David L. Downe, and Janet Welsh Brown, *Global Environmental Politics*, Fourth Edition (Cambridge: Cambridge University Press, 2006), pp. 42–43.

23. John Rodman, "Paradigm Change in Political Science: An Ecological Perspective," *American Behavioral Scientist*, 24:1, 1980, pp. 49–53.

24. Ibid., pp. 51–53.

25. Larry Diamond. *Developing Democracy* (Baltimore: Johns Hopkins University Press, 1999), pp. 10–15.

26. Joseph Migdal. "Studying the State" in *Comparative Politics, Rationality, Culture and Structure*, Mark Irving Lichbach and Alan S. Zuckerman (eds.), pp. 208–236.

27. Ibid., pp. 228–230.

28. Ibid.

29. Kubasek, Nancy K. and Gary S. Silverman. *Environmental Law*, Fourth Edition (Upper Saddle River: Pearson, 2002); Water Policy Control, pp. 189–190.

30. Ibid.

31. Ibid.

32. Ibid.

33. Ibid.

34. Ibid.

BIBLIOGRAPHY

Alexander v. Sandoval, 532 U.S. 275 (2001) in Alexander v. Sandoval: 532 U.S. (2001): Justia US Supreme Court Reporter, last date accessed 22-3-2021.

Bullard, Robert D. "Unequal Environmental Protection: Incorporating Environmental Justice in Decision Making" in *Worst Things First* (Adam M. Finkel and Dominic Goldings eds.), London: Routledge, 1994.

Chasek, Pamela, David Downe, and Janet Welsh Brown. *Global Environmental Politics*, Fourth Edition, Cambridge: Cambridge University Press, 2006.

Diamond, Larry. *Developing Democracy*, Baltimore: The Johns Hopkins University Press, 1999.

Gordon, Frederick D. *The Damming of the Presidency: How Environmental Factors Impact a Political Campaign.* Lanham: Rowman and Littlefield, 2020.

Kosinsky, David. Ed. *Failed Promises,* Cambridge: MIT Press, 2015.

Kubasek, Nancy K. and Gary S. Silverman. *Environmental Law*, Fourth Edition, Upper Saddle River: Pearson, 2002.

Kuehn, Robert R. "A Taxonomy of Environmental Justice," The University of Alabama School of Law, *Environmental Law Reporter*, vol. 30 (2000), p. 10681.

Migdal, Joseph. "Studying the State" in *Comparative Politics: Rationality, Culture and Structure* eds. Mark Irving Lichbach and Alan S. Zuckerman, Cambridge: Cambridge, University Press, 1999.

Rodman, John. "Paradigm Change in Political Science: An Ecological Perspective," *American Behavioral Scientist*, 24:1, 1980.

Scott, Shirley. *International Law in World Politics: An Introduction*, Second Edition. Boulder: Rienner, 2010.
Yokota, Yozo. "International Justice and the Global Environment" in *Journal of International Affairs*, 52:2, 1999.

Chapter Four

Moving Michigan Towards Just Building, Housing, Energy Production, Distribution and Storage

EGLE Council on Climate Solutions and the Limits of (Neo)/Liberalism

Michael J. Lorr

From working to teaching, to shopping, to caring for ourselves and others, the pandemic has required everyone to rethink their daily lives. It also called attention to how unready and unwilling society is to deal with a global catastrophe. While still running triage on a resurging pandemic, life during the COVID-19 outbreak has highlighted the depths of racial, class, gender, and socio-spatial marginalization in the United States and the interlocking systems of global inequality faced today. Simply put, 1 million dead Americans and 6.5 million dead globally is still not enough to inspire humanity to take concerted action to effectively address the pandemic let alone the other current and near future climate change caused calamities. In the face of these extreme conditions, energy justice may provide a step in the right direction.

The Biden administration reengaged the United States in the Paris Climate Agreement on January 20, 2021 to reduce greenhouse gas (GHG) emissions and in the state of Michigan Governor Gretchen Whitmer was asking for recommendations to make Michigan carbon free by 2030. This chapter will look at that state-wide process from April to September 2021, as it highlights some interesting ways in which government, non-profits, and industry have used the pandemic shifts to virtual meetings to at least make it appear that more people can have a say in policy proposals and decision-making. Specifically, this chapter is comprised of participant observation of the Michigan Environment, Great Lakes, and Energy (EGLE) Council on Climate Solutions open working subgroups of (1) building and housing and (2) energy production, distribution and storage. These were two of six working groups tasked with creating recommendations and proposals for the Governor to decarbonize Michigan by 2030. This chapter focuses on the limits of this type of (neo)/liberal process and how energy justice and other "justice" forward perspectives may be informing these processes to what end.

LITERATURE

The scientific evidence that human-instigated climate change is a real threat is pretty much unequivocal at this point. Human actions are the main cause of this catastrophic problem. Despite the existing evidence, fossil fuel consumption and global energy demand contune to rise as if there is no problem and no limit to human consumption. The twin crises of failing or at least faltering democracies and failing health as evidenced by the January 6 coup attempt in the United States and the global COVID-19 pandemic have only intensified the fragility of humanity on Earth. If there is any upside or silver lining to these twin crises, it may be how in particular the quarantine phase of COVID-19 had immediate impacts on health, wellbeing, energy use, and GHG emissions—for example, shifting to at-home and virtual teaching, learning, and work arrangements radically reduced the fossil fuel pollution from commuting for a time and there was also a brief moment where it seemed like people were actually concerned with mental health and the prevention of burn out, the grind, and other maladaptive consequences of contemporary capitalism. The question that remains now in the aftermath of the destabilization of the peaceful transfer of power in the United States and the pandemic and climate crisis, is if humanity has learned any lessons from these events. Perspectives like "energy justice" argue that all current energy systems from legal, political, and philosophical justifications need to be overhauled to give primacy to the idea of justice as energy policy systems are often detached from energy justice and ethics concepts (Jenkins et al. 2018). Salter et al. (2018) argue that there is a growing energy law framework, bending toward this direction of justice that is developing rapidly in the U.S. and worldwide, especially in environmental law and natural resources law. Energy justice, a relatively new concept compared to environmental justice, but similar to concepts like 'just' sustainability can be characterized as a tool for policymaking processes that seeks to identify when and where injustices occur and how to identify them and maybe even address them (Agyeman et al. 2003).

At this point it almost feels disingenuous to keep on saying that climate change is an existential threat to human society as we know it, but powerful people like the World Bank President David Malpass still regularly deny that humans are the cause of the current and ongoing climate catastrophe. One of the paradoxes of the climate debate, which has stymied thinking and policy-making on this issue, is do all humans cause climate damage equally or does in reality a minority of wealthy nations, corporations, political and moneyed elites, and consumers hold the lion's share of responsibility for this crisis? Humanity's weak mitigation of climate change over the past few decades has inspired many attempts to motivate action by identifying this exceptional climate emergency and its existential threat to humanity (Oels 2013, von Lucke 2014). One approach has been to characterize the unprecedented scale of anthropogenic impacts on the environment as a distinct geological era. But this approach of naming the current age the "Anthropocene" promotes the idea that all humans are equally responsible for the current sixth extinction event (Kolbert 2014, Malm and Hornborg 2014, Davis and Todd 2017). The Anthropocene approach effectively elides social inequalities and again effectively ignores the fact that a minority of wealthy nations, corporations, political and moneyed elites, and consumers are overwhelmingly responsible for the destruction of the planet (Davis et al. 2019).

Energy and climate justice, and 'just' sustainability approaches focus on addressing unequal climate impacts by using redistributive schemes like reparations, loss and damage, risk transfer and polluter-pays policies like carbon taxes (Page 2008, Burkett 2009, Grasso 2010). But these tools of financial compensation only work when wrongdoers and victims are easily identified (Barnett 2020). The conundrum of this climate change paradox is evident in the intense debates among parties to the United Nations Framework Convention on Climate Change over terminologies such as "loss and damage" and "liability and compensation" and in the United States over whether it should be party to the United Nations or the Paris Climate Agreement in the first place—the answers change when administrations do. Povinelli (2016), Maboloc (2020), and McLaughlin (2020) argue that there are definite limits imposed on climate justice approaches by liberalism. It is unlikely that energy and climate justice, and/or 'just' sustainability can come to fruition in a world system that relies on the exploitation of places, animals, and peoples not recognized as fully human in order to function. Aside from questions of practicality, some decolonial thinkers are beginning to challenge the premise that climate injustices can be addressed through legal means in the first place (Anghie 2007). While the above-mentioned justice forward perspectives, climate and energy justice and 'just' sustainability among others, are attempts to address this paradox and inform more effective solutions to the climate crisis what are the limits to these approaches?

As Harvey (2005) and Klein (2007) have argued before, one of the main problems of the contemporary neoliberal era at least politically in the United States is that both parties have more in common with corporate elite interests than average U.S. voters. One reason for this is that during the shift from the New Deal "Fix," which regularized the idea of a social safety net for people to the Neoliberal "Fix," which regularized a social safety net for corporations, government abdicated its responsibility to help people or subsidize the development of its citizenry and now instead more often sees its responsibility as to help business and more often to subsidize the risk of large corporations. This abdication of responsibility to its citizenry by both parties is compounded by the post-democratic swing toward outright authoritarianism at least since the advent of the Trump administration in 2016, although the precursors have been present arguably at least since Ronald Reagan, if not earlier with the advent of the counterintelligence program (cointelpro) in the 1960s (Crouch 2020). The United States has heavily subsidized the fossil fuel industry over the years, and it has only been very recently that a shift toward renewables with the Obama administration has begun. Unfortunately, some of the strides in the renewable's direction at least symbolically with the signing of the Paris Climate Accord during Obama, were at least symbolically taken back during the Trump years and now they have been reinstated by the Biden administration.

While both parties have a relatively dismal record on environmental action, with the current Democrats being at least symbolically better, both parties have also been mostly ineffectual on issues of environmental justice and decolonization. Agyeman et al. (2003) have argued that in order for environmental sustainability to mean anything useful, justice—in the form of racial, class-based, and gender equity—needs to be front and center. And further Tuck and Yang (2019) argue that justice needs to include repatriating unceded land, restoring tribal sovereignty, abolishing extractive systems, and repairing past harms; otherwise, justice for settlers will only perpetuate the oppression and domination of the land, animals, and indigenous peoples and decolonization will become a useless metaphor.

During the change of administrations from Trump to Biden (alarming in its own right for the January 6 insurrectionist coup attempt of Trump loyalists which interfered with the peaceful transfer of power in the United States), the Biden administration was attempting to take more aggressive action on the climate issue. In the state of Michigan, this desire of the larger administration had Governor Gretchen Whitmer soliciting recommendations from the state's Department of Environment, Great Lakes, and Energy (EGLE) to make Michigan's economy carbon free by 2030 (not to spoil it, but the revised goal is now officially 2050). This process offers some insights into the limits of neoliberalism addressing social and environmental problems as well as the potential promise of more radical approaches like environmental justice and decolonization. This chapter intends to document how one state in the United States (Michigan) is attempting to address these issues. Two questions drive this work: how can states bring more stakeholders to the table to effectively address these issues and how can community participants effectively get their issues addressed? I see this research intersecting with two literatures: one on the limits of neoliberalism addressing social and environmental problems and the second on the potential promise of more radical approaches like environmental justice and decolonization.

METHODS

As a long-time sociological participant observer, I completed this study by becoming aware of an opportunity to serve on Michigan's Department of Environment, Great Lakes, and Energy (EGLE) Council on Climate Solutions open working subgroups of (1) building and housing and (2) energy production, distribution and storage through the environmental justice work I contribute to with an environmental non-profit that primarily works on water justice issues called West Michigan Environmental Action Council (WMEAC). After responding to a call for participants via email, I was invited to participate and contribute to EGLE's Council on Climate Solutions working subgroups. We were tasked with delivering policy recommendations to the Governor of Michigan in order to decarbonize Michigan's economy by 2030. The Council's subgroups met virtually on Zoom between April and September 2021. I participated in 12 of the 14 total meetings. I studied and contributed to meeting materials and online "sharesites" for the two working groups. During these twelve meetings I participated as both a representative of WMEAC and as an observer studying the process. In the first meetings there were close to 300 participants Zooming in. During the subgroup meetings there were closer to 30 participants. During the final larger group meetings there were closer to 100 participants.

Mindful of my role as participant and observer, it was interesting to react on the one hand as a representative of my community and the non-profit I work with and on the other hand as an academic. Experiencing the meetings as a representative of my community and the nonprofit I work with, I felt like I had a voice in some important environmental policy discussions and deliberations that may ameliorate some of the climate catastrophe we are currently experiencing. Experiencing the meeting as an academic sociologist, I was struck positively by the possibility for many voices to be heard in the setting of online meetings via Zoom, but negatively I was struck by how quickly critical participants fell out of the meetings and that the voices that really had sway and continued participation were representatives of energy companies or consultants of industries who would either benefit or be harmed from policy rather than citizens of the communities that these policies may help or harm. The next section offers findings from this short-term participant observation study keeping in mind the limits of neoliberalism and the promise of approaches like energy justice to answer how can states bring more stakeholders to the table to effectively address these social environmental issues and how can community participants effectively get their issues addressed?

FINDINGS

When I responded to the call for participants to contribute to Michigan's Department of Environment, Great Lakes, and Energy (EGLE) Council on Climate Solutions, it was right when the snow was starting to make way for spring in 2021 and almost a year and a half into radically altered pandemic life routines. I thought that this Council's virtual approach was novel for soliciting input on new environmental policy and also necessary to safely convene large groups of people while preventing them from physically coming together. These online virtual meetings approach perhaps coupled with the pandemic inspired desire to be social with people while social distancing enabled the Council to hear and engage with multiple perspectives and stakeholder's ideas on how to decarbonize Michigan's economy by 2030.

APRIL 2021: THE SET UP

During the first virtual Zoom meeting there were many people attending with a diversity of perspectives. These first sessions were run almost the same way I was running my own virtual college classrooms at the time with blocks of information delivery, lecture, presentation and regular use of randomized virtual breakout rooms to encourage deeper discussions and deliberations. The approximately 300 participants first heard from the director of EGLE to set the stage for how this process was going to work. The Council on Climate Solutions consists of 14 Michigan residents appointed by the governor to represent a range of sectors, experiences, and expertise relevant to climate issues. Additionally, the Council includes the directors (or their designees) of EGLE and the Michigan Departments of Agriculture and Rural Development (MDARD), Labor and Economic Opportunity (LEO), Michigan Economic Development Corporation (MEDC), Natural Resources (DNR), Transportation (MDOT), and Health and Human Services (DHHS), as well as the State Treasurer and the Chair of the Michigan Public Service

Commission (MPSC). At this first meeting and others that took place in that first month, we heard from experts in fields pertaining to different aspects of attempting to decarbonize the economy and from racial and environmental justice experts to set the stage for our tasks in the six working subgroups: 1) energy, transmission distribution and storage, 2) transportation and mobility, 3) energy intensive industries, 4) buildings and housing, 5) natural and working lands and forest products, and 6) materials decarbonization. I became a member/participant of the 1) energy, transmission distribution and storage and 4) buildings and housing subgroups.

JUNE–AUGUST 2021: CREATING RECOMMENDATIONS TO DECARBONIZE MICHIGAN BY 2030

After the first month of the general set up meetings we then broke into our working subgroups for the bulk of our time together where we also heard from experts on the more specific issues pertaining to the subgroup. What struck me most at this point was how similar to my college classrooms they were in that there was a kind of following the experts or halo effect/fallacy in the large groups and then at the end or middle of the meetings when we were broken into smaller break out rooms of 4–5 people then the critiques would really start flowing. Many of the main presentations and the experts who had the floor were from industry pushing particular technological fixes with a combo of tax breaks or purchasing incentives for implementation. For example, in the buildings and housing subgroups experts were aggressively pushing for the elimination of gas stoves in houses in favor of electric stoves as well as aggressively pushing for schemes that would have home or building owners buying and installing heat pump technologies instead of the currently conventional gas and electric dependent systems. This was novel to me in the sense that I just came across an article in *Mother Jones* that made the argument that natural gas is the number one way we pollute our own homes and that the natural gas lobby is resorting to hiring influencers to gin up support among millennials to see natural gas stoves as cool and the height of luxury (Leder 2021). Still, this seemed like a very specific and a possibly in the weeds point of contention for the workgroup deliberations and what really struck me here was how there was no real solicitation or study of what citizens actually want in the way of building, housing, energy production, transmission, distribution and storage. The process was more of a presentation of what different industries would like to be incentivized to offer citizens instead of the citizenry telling or demanding what they actually want. When alternate ideas were offered that did not fit this format like an idea of decentralizing the power grid and more solar powered homes with an end goal of getting rid of the larger power and electric companies, there was shock that participants were suggesting these kinds of solutions that would in the end dissolve the large-scale power companies. There was also shock and derision when participants suggested reverting these utilities to public ownership. The most radical they allowed was for a cooperative model that is used in the Upper Peninsula, but as in keeping that model up there and not as a desire to have that model come to the rest of the state. Maybe this seeming pro-business, pro-fossil fuel bent was just an aspect of the kind of people who dominate the housing and energy sectors in the state of Michigan and did not represent the ideas in the other working groups. For example, I found out later that the natural and working lands and forest products subgroup regularly brought up and centered equity and inclusion of perspectives at every meeting. Maybe the housing and energy sector workgroups lacked a diversity of backgrounds that other groups may have had. Maybe both the buildings and housing and the energy, transmission distribution and storage subgroups trended older, whiter, male and as more heavily industry based or influenced than some of the other groups. For example, many of the participants in the buildings and housing subgroup were contractors and building owners and, in the energy, transmission distribution and storage subgroup, participants were mainly utilities representatives and wonky energy engineers not community members, citizens, or activists. As participation dwindled in the later summer months and with the length and span of the meetings, I started to get the sense that most of the people continuing to participate were paid to be there as a part of their normal job responsibilities, that is gas and energy company people and government people, not many actual citizens or community members. So, what started as a potentially inclusive process via virtual online meetings seemed to shift back to the more typical industry policy insiders via attrition through the time of the process and the centering of technological fixes as opposed to social justice-oriented ideas.

SEPTEMBER 2021: HUNDREDS OF RECOMMENDATIONS BECOME 5 PER SUBGROUP

During the month of September all of the working subgroups were tasked with writing up all of their recommendations to be narrowed down to a top 5 and then submitted to the governor appointed Council on Climate Solutions for further public comment, study, and possible implementation. For the buildings and housing subgroup this meant taking 73 recommendations and editing them down to an actionable top 5. In the end, the broad recommendations from the council that loosely matched up with each working subgroup were to: 1) Commit to environmental justice and pursue a just transition; 2)

Clean the energy grid; 3) Electrify vehicles and increase public transit; 4) Repair and decarbonize homes and business; 5) Drive clean innovation in industry; and 6) Protect Michigan's land and water. Under each of these broad recommendations are the top 5 or 6 of each subgroup's ideas. For example, "the energy, transmission, distribution, and storage" working subgroup became the "clean the energy grid" recommendation with these key strategies from the group: clean energy resources, holistic and integrated energy system planning, affordability and energy burden, and siting and energy storage. Whereas the "buildings and homes" working subgroup became the "repair and decarbonize homes and business" recommendation with these key strategies from the group: heating Michigan homes and businesses, clean financing opportunities, building codes and climate readiness, energy and waste reduction programs, incentives for energy efficient appliances, and evaluate gas system regulatory policies. Before turning to the discussion section, it is important to mention that around this time in late August/early September we were told that while 2030 was still the stretch goal to decarbonize Michigan's economy the new target date was now officially 2050.

DISCUSSION

In the work groups that I was privy to, it was apparent that critical perspectives made officials and industry representatives squirm. While looking at the finalized recommendations it seems like equity was a virtue signal of this policy writing by committee even if in fact equity was not. When activists in the energy, transmission, distribution, and storage subgroup put forward a Community Ownership of Power Administration or (COPA) proposal like the kind in some conceptions of a Green New Deal, which calls for an end of investor owned utilities because they prioritize money-making over community needs in favor of a COPA resembling President Roosevelt's New Deal era Rural Electrification Administration, but centering racial and economic justice, it was considered interesting but then put aside in favor of tweaking the current investor owned utility scheme.

There are many interesting first steps in the outcome of these deliberations, which is titled: *MI Healthy Climate Plan*. Its goal of 2050 for full decarbonization may support the idea that there are at least some inroads being made by energy and environmental justice approaches on how policy gets made and who it might help or harm (access and download the finalized plan here: https://www.michigan.gov/egle/about/organization/climate-and-energy/mi-healthy-climate-plan). However, it seems that the influence of energy justice ideas may still be at the level of the kind that remains virtue signaling versus actual substantive change toward real justice, or real representation, or even more radically real decolonization and abolition of harm or reparations for past harms. Some of the bench marks set by the plan are bold if they are enacted like 40% of state funding for climate initiatives need to benefit Michigan's disadvantaged communities, generating 60% of the state's electricity from renewables by 2030, limiting the cost of power and heating homes to no more the 6% of annual income for low-income households, increase public transit and electric vehicle usage by 15% every year, reduce home and business heating related emissions by 17% by 2030, triple the states recycling rate to 45% and cut food waste in half by 2030, and protect 30% of Michigan's land and water by 2030 for natural carbon capture.

It is unclear if bench marks like these will alter the inequality, injustice, and steady environmental erosion experienced currently. It is clear that the groups convened during this policy recommendations process were at least aware of and some were influenced by energy and environmental justice ideas, but less aware of or influenced by ideas of decolonization, abolition of harms, or reparations for past harms. This short-term participant observation study does show that states could bring more stakeholders to the table to effectively address these social environmental issues by using more of the tools they have available like Zoom and maybe creating some other types of participation incentives or payments. This council was successful at getting around 300 community members, industry experts, and policy makers together by using Zoom as the virtual meeting space. Online technology, while still exclusive because of the potential to exclude due to the cost of a computer and internet access was possibly made less exclusive by the normalization of online gatherings due to the COVID-19 pandemic. On the other hand, at least 2/3 of the original participants dropped out of the process from the beginning to the completion of recommendations. States should pay those who are attending just as functionaries for energy or engineering firms are paid if states want to try to reduce the dramatic participation melt that occurred here and probably led to less progressive and more middle of the road policy suggestions.

In terms of community participants effectively getting their issues addressed, policy writing by committee in which 72 recommendations are combined or reduced to 5 felt like only so much community data harvesting or listening sessions where the purpose is to make participants feel like they are heard, but to then have the actual strength of the recommendations trend to predetermined ideas. In the working groups I was a part of this was like how community based or decentralized power systems ideas that could free citizens from monopolistic investor-owned utilities were massaged into proposals for electric stoves and heat pumps purchasing incentives or similarly when COPA was proposed as a method to reduce emissions were instead shifted toward highly technocratic ideas of carbon capture technologies.

More broadly, in a society in which democracy is rarely actually practiced by its citizenry besides for perfunctory votes for representatives a couple of times a year and is actively under threat, the kind and level of participation asked for by this council has probably ended up with recommendations that everyday voters would need more explanation and deliberation. But the process is a start and it will be interesting to see how well it is implemented or stymied by future administrations.

CONCLUSION

This short participant observation study shows the possibility of how virtual meeting technology like Zoom could bring lots of people together to talk about, debate, and create more citizen informed policy. Unfortunately, the usual suspects in government and industry eventually drowned out or wore down the voices of community, citizens, and more critical perspectives in essence pointing out the limits to (neo)/liberal approaches and the traction or lack of traction energy justice and more critical approaches have on the process. While Governor Gretchen Whitmer's *MI Healthy Climate Plan* may put Michigan on the path toward achieving 100% carbon neutrality statewide by 2050 and take some important action to address the climate crisis, it remains to be seen if the prioritizing the health of Michigan communities now and protecting future generations from climate change will be enough or amount to a too little, too middle-of-the-road, too late variety. In the coming years, future studies should attempt to assess how adequately the goals of *MI Healthy Climate Plan* have been implemented and also attempt to assess if these plans were aggressive enough in the first place. I would hypothesize that while these plans seem bold that they are actually very much in line with past examples of neoliberal governance and while in many ways this healthy climate plan appears to be bold it is also a case study in the limits of neoliberal strategies to fix climate problems. The ongoing climate catastrophe was created by the state's unwillingness to meaningfully study, engage, and cultivate its citizenry and worse its inability to understand its colonial nature harming those peoples that came before its existence.

BIBLIOGRAPHY

Agyeman, Julian, Bullard, R., & Evans, B., eds. 2003. *Just Sustainabilities: Development in an Unequal World.* Cambridge: MIT Press.

Anghie, A. 2007. Imperialism, Sovereignty and the Making of International Law. Cambridge, UK: Cambridge University Press.

Barnett, J. 2020. “Global Environmental Change II: Political Economies of Vulnerability to Climate Change.” Progress in Human Geography, 44(6), 1172–1184.

Burkett, M. 2009. “Climate Reparations.” Melbourne Journal of International Law, 10, 509–542.

Crouch, Colin. 2020. *Post-Democracy After the Crises*. Cambridge: Polity Press.

Davis, H., & Todd, Z. 2017. On the Importance of a Date, or Decolonizing the Anthropocene. ACME (Vol. 16, pp. 761–780).

Davis, J., Moulton, A. A., Van Sant, L., & Williams, B. 2019. “Anthropocene, Capitalocene, Plantationocene?: A Manifesto for Ecological Justice in an Age of Global Crises.” Geography Compass, 13, 1–15.

Grasso, M. 2010. “An Ethical Approach to Climate Adaptation Finance.” Global Environmental Change, 20(1), 74–81.

Harvey, David. 2005. *A Brief History of Neoliberalism*. New York: Oxford University Press.

Jenkins, K., Sovacool, B. K., & McCauley, D. (2018). “Humanizing Sociotechnical Systems through Energy Justice: New Conceptual Frameworks for Global Transformative Change.” *Energy Policy, 117*, 66–74.

Klein, Naomi. 2007. *The Shock Doctrine: The Rise of Disaster Capitalism*. New York: Picador.

Kolbert, E. 2014. *The Sixth Extinction: An Unnatural History*. New York: Picador.

Leber, Rebecca. 2021. “Gaslit: How the Fossil Fuel Industry Convinced Americans to Love Their Toxic Stoves.” *Mother Jones* (July/August), 42–47.

Lucke, von, F., Wellmann, Z., & Diez, T. 2014. “What’s at Stake in Securitising Climate Change? Towards a Differentiated Approach.” Geopolitics, 19(4), 857–884.

Maboloc, C.R. 2020. “Liberal Environmentalism and Global Climate Justice.” Eubios Journal of Asian and International Bioethics, 30(2), 51–56.

Malm, A., & Hornborg, A. 2014. “The Geology of Mankind? A Critique of the Anthropocene Narrative.” The Anthropocene Review, 1(1), 62–69.

McLaughlin, A. 2020. “The Limit of Climate Justice: Unfair Sacrifice and Aggregate Harm.” Critical Review of International Social and Political Philosophy, DOI: 10.1080/13698230.2020.1786306

Oels, A. 2013. “Rendering Climate Change Governable by Risk: From Probability to Contingency.” Geoforum, 45(C), 17–29.

Page, E. A. 2008. “Distributing the Burdens of Climate Change.” Environmental Politics, 17(4), 556–575.

Povinelli, E.A. 2016. Geontologies: A Requiem to Late Liberalism. Durham: Duke University Press.

Salter, Raya, Gonzalez, Carmen, & Warner, Elizabeth. 2018. *Energy Justice: US and International Perspectives*. Cheltenham: Edward Elgar Publishing.

Tuck, Eve, & Wayne Yang, K. 2012. “Decolonization Is Not a Metaphor.” *Decolonization: Indigeneity, Education, & Society*, *1*(1), 1–40.

Chapter Five

Sustainable Electric Energy Development in Bahia

Lucigleide Nery Nascimento

Energy is a necessary resource in the current global societies; the "access to affordable, reliable, sustainable and modern energy" is expressed in the 2030 Agenda (2015–2030) for sustainable development (The United Nations, 2019). The use of electric energy has been employed as a proxy to assess the improvement in the material condition of the population. According to the Universal Declaration of Human Rights (The United Nations, 1948), "everyone has the right to a standard of living adequate for the health and well-being of himself and of his family." In Brazil, as in Bahia, to ameliorate people's lives, it has been an increase in the number of individuals and households using electric energy. Indeed, the access to electricity has been seen as almost universalized in that Brazilian state. But urban areas have presented a more favorable situation than rural zones (Pesquisa Nacional por Amostra de Domicílios Contínua, 2019).

A little over 30% of all types of energy supplied in Bahia comes from renewable sources (Secretaria de Infraestrutura do Estado da Bahia, 2018; Empresa de Pesquisa Energética, 2023a, 2023b, 2023c, 2023d). In the demand side, the transportation, the industrial and residential sectors have the highest final energy consumption of Bahian energy grid (Empresa de Pesquisa Energética, 2023d). Regarding electricity generation, historically, hydropower has been the major type, but wind generation has been a growing trend in the last decades. In 2022, the industrial sector stood out as the segment with the highest usage of electricity in Bahia, followed by, in this order, residential and commercial divisions (Empresa de Pesquisa Energética, 2023a).

Nonetheless, the production and use of energy give rise to innumerable consequences or externalities in Brazil, Bahia and elsewhere (Gleick, 2009;

Nascimento, 2010). Worldwide, the dependency upon conventional energy and its production increased emissions of greenhouse gases and contributed to climate change and its effects. According to the State of the Global Climate 2022 (World Meteorological Organization, 2023, 1), "increasing levels of greenhouse gases in the atmosphere due to human activities have been the major driver of climate change since at least the industrial revolution." Climate change and its effects continued to advance in 2022 (World Meteorological Organization, 2023).[1] Then, for that reason, the societies need to follow a net-zero emission pathway.

Large-scale hydropower infrastructures and their management have caused social and environmental impacts in Brazil, altering from landscape to way of living: forced displacements, river fragmentation and impaired fishery (Nascimento, 2010). In conjunction with climate variability and change impact water availability and quality, such as its spatial and temporal distribution (World Meteorological Organization, 2022). Nations, such as Brazil, dependent upon hydropower generation, need to diversify sources of electricity.

Technology improvement has favored the expansion of new forms of generation of energy such as renewable, low-carbon or emissions-free; and has made the transition towards renewable energy cheaper and more accessible to consumers (World Meteorological Organization, 2023). Governmental incentives, via policy instruments, have been necessary. In Brazil, the Programa de Incentivo às Fontes Alternativas de Energia Elétrica—PROINFA (The Alternative Source of Electric Energy Incentive Program) (Law No. 10,438, 2002) instigated the use of new forms of renewable energy.

This study investigates the trends in sustainable electric energy development in Bahia in the last decade and into the near future. This work then adds to the literature on policy, renewable energy development, environmental and socio-economic studies. The chapter consists of 5 sections. The first one is this introduction. The second one provides a brief review of the literature on renewable energy and policy instruments. The third section describes the methodology, the Policy Sciences Analytic Framework (PSAF). The fourth part details the case study. The last segment is the conclusion.

LITERATURE REVIEW

Renewable Energy

According to the United Nations, "[r]enewable energy is energy derived from natural sources that are replenished at a higher rate than they are consumed. Sunlight and wind, for example, are such sources that are constantly being replenished" (The United Nations, 2023). Since renewable energy resources (hydro, wind, solar) are more spatially distributed than fossil fuel reserves (coal, natural gas, oil) (Overland, Juraev, Vakulchuk, 2022), it favors decentralized power generation. This characteristic has influence upon world's geopolitics due to changes in the way energy resources have been used and controlled: reliance on fossil fuels.

According to McKenna et al. (2022), potential in renewable energy has a broad possibility of meanings. For instance, it is physical when denotates energy content of wind and it is geographical if it refers to the available area for wind turbine installation considering the constraints. "The technical potential corresponds to the wind power generated within the geographical potential. It considers constraints such as wind turbine characteristics, wind farm array losses and electrical conversion losses" (McKenna et al., 2022, 661). Economic potential focuses upon profitability (McKenna et al., 2022).

Wind and solar power are intermittent. Production varies due to seasonal and diurnal patterns. But, even so, renewable energy is no longer an alternative group of energy, it is now mainstream in the electricity production sector (REN21, 2023). Worldwide, it has been an increase in installed capacity, investments and households' access to distribution's systems (REN21, 2023). Wind and solar power are more cost effective than coal-fired power (REN21, 2023).

Even so, challenges still exist for the RE's sectors. Traditional sectors maintain governmental support. For example, investments as well as subsidies are applied in the new fossil fuel production (REN21, 2023). Despite technological advances, as with any machinery, there is a decline in performance with age of the wind turbine (Staffell, Green, 2014). In the case of wind power generation, it could result from gradual deterioration of a rotating machine, reduction of the efficiency and difficulties to reestablish operations after fails. The energy produced reduces overtime and has an impact upon the economic life of the equipment. The degradation is not insignificant but takes place also with conventional generation technologies and possibly has a low impact on costs of wind energy, but should be considered (Staffell, Green, 2014).

Solar power presents issues too. The energy results from solar radiation. "Solar resource distribution is non-uniform and intermittent, and highly dependent on external factors such as geographic location, different meteorological parameters, and weather conditions" (Hao et al., 2022, 693). The dust deposition on panels has also influence, reduces power generation. The decision of the location of the site is important for local ecosystems, since solar panel occupies large areas with shadow.

Policy Instruments

Several types of policies and instruments have been experienced, including in the energy field. "Mission-oriented" focuses upon specific objectives, setting directions to achieve them (Mazzucatto, 2018). UN climate action seeks the net-zero commitment: to cut greenhouse gas emissions to as close to zero as possible; and to achieve this, renewable energy is crucial to address climate change. The demand-oriented approach focuses on increasing demand of users and encouraging suppliers to meet such needs, such as with guaranteed tariffs for RE (Organization for Economic Cooperation and Development, 2011). Furthermore, investment-oriented policy via access to finance, such as loans, guarantees and other instruments, can play an import role and contribute to renewable energy development, since the sector "requires significant upfront investment" (Wishlade, Michie, Vernon, 2017, 25). In the case of renewable energy, a mix of policy approaches has been applied.

Joint efforts (public, private and third sector) are important in the creation of new markets (Mazzucato, 2018), here for instance, renewable energy ones. But examples show public policy's importance. European countries, for example, have established targets to use 100% of renewable electric energy (REN21, 2023). Policy instruments, technological change and diffusion have spurred the development of RE (Zhou et al., 2021). Economics is an important variable as well (Zhou et al., 2021).

THE POLICY SCIENCES ANALYTIC FRAMEWORK (PSAF)

The PSAF is a tool for policy analysis (Lasswell, 1965). Developed by Harold Lasswell in the 1930s, PSAF's three major dimensions are: problem orientation, context or social process and decision process (Clark, 2002). The instrument aims to improve knowledge of and in policy (Lasswell, 1971). Policy methodologies focus on a problem and are based upon the whole policy and social processes. Under such approaches, prescriptions are not created apart from the setting where the issue of concern takes place (Lasswell, 1970). "[P]olicy approaches tend toward contextuality in place of fragmentation, and toward a problem-oriented, not problem-blind, orientation" (Lasswell 1970, 6).

This chapter applied the Policy Sciences Analytic Framework (PSAF) to guide the understanding and to assess the case of the renewable energy development in Bahia, Brazil (Lasswell, 1971; Clark, 2002). The case is complex and involves social, economic and environmental aspects. This research is based upon the collection and the critical review of empirical data, relevant legal statutes, and secondary literature (Maxwell, 1996).

Problem Orientation and Major Prescriptions

The recent expansion of renewable energy production in Brazil has been driven by policy for decentralization of generation and diversification of sources. The decision process involving alternative sources was a reactive measure to address hydropower energy shortages in 2001 and 2002. The *Programa de Incentivo às Fontes Alternativas de Energia Elétrica*—PROINFA (The Alternative Source of Electric Energy Incentive Program) has existed since 2002 (Law No. 10,438, 2002) (Brasil, 2002). The general objective was to increase the share of electrical energy from renewable sources supplied by independent power producers. The goal of the first phase of the program was the implementation of 3300 MW of installed capacity of alternative sources, equally, from wind, small-scale hydropower plants and biomass (Brasil, 2002). The program set the limit to the use of national equipment and services in the provision of electrical energy from alternative basis.

The Brazilian Federal Government has provided different types of policy instruments that has influenced the current production of RE in that nation. It has set targets, such as in the first phase of the PROINFA program; applied the Feed-in Tariff, buying electricity at a specific price for a long-term period; provided tax cuts and exemptions; and adjusted policies as well, due to the initial inability of the domestic market to provide technology.

Law No 10,848/2004 established the new model of electricity commercialization (regulated or free market) (BRASIL, 2004). Through energy auctions, the federal government has played an important role in the expansion of renewable generation (Pinheiro, Arariba, 2017). The federal government regulates the commercialization of electric energy and the process of concession and authorization for the generation of electric energy (Decree No 5,163/2004) (Brasil, 2004). Auction is an instrument applied to contract installed capacity of electricity to supply the market. The tool has been used to influence tariff, select source aiming modernization and diversification of the Brazilian electric energy grid. The federal government bought installed capacity in auctions at prices for the specific source above an established price floor for a long-term period. In 2004, the first public call to contract energy from renewable sources took place and more than one-third of the total agreed was from wind power (Dutra, Szklo, 2008). In 2009, the federal government held the first specific public auction to contract reserve of wind energy to meet national demand (Ministério de Minas e Energia do Brasil, 2010). In 2014, a historical public auction exclusively contracted installed capacity of solar power for the regulated market (Empresa de Pesquisa Energética, 2014). Solar energy has a more decentralized nature (Shimomaebara, Peyerl, 2021).

Financial support has existed via funding programs from the Brazilian Socio-Economic Development Bank (Banco Nacional de Desenvolvimento Econômico e Social)—the BNDES and the Superintendence for the Development of the Northeast (Superintendência do Desenvolvimento do Nordeste)—SUDENE; and tax breaks for renewable energy equipment/parts (Nascimento, 2019). But the companies providing RE need to overcome challenges, such as lack of national technologies, land ownership issues, workforce gap and difficulties to connect the powerplant to the national grid.

International Context

The consequences of global warming and climate change and the claim to reduce or eliminate GHG emissions stimulate the use of renewable energy. Brazil has ratified the Paris Agreement on Climate Change committing to cut greenhouse gas emissions. As a member of the United Nations and towards the achievement of the Sustainable Development Goals (SDGs), the country works towards the diversification of the energy sources increasing the proportion of RE and contributing for the reduction of GHG emissions.

The scale of production and the development of new technologies/innovations have influenced prices of both energy from solar and wind sources and equipment (Santos, Torres, 2017; Shimomaebara, Peyerl, 2021). Santos and Torres (2017), for example, acknowledge China's importance in those matters in the international scenario. Initially, in Brazil, the inexistence of national technology was a challenge (cost). But, today, the increase in the size of the new sets of equipment has spurred generation's capacity: "taller turbines and larger rotor diameters" (The United Nations, 2023).

The Socioeconomic Perspective

Among the social benefits linked to renewable energy are job creation and local economic development. In her studies about wind energy in Brazil, Traldi (2017) states that, the authors in Brazilian literature in favor of that argument mention the new opportunities during the implementation phase (more numerous, but temporary occupations) and the stage of operation/ maintenance stage (fewer jobs, but permanent positions). In addition, the multiplier effects occur in the local economy, that means, additional benefits exist in the local business where the income from work and land rent for wind turbines is spent.

Authors on the other side of the debate argue that the benefits in jobs creations are of minor proportions (more during the construction phase). There is an increase in local inflation due to the temporary demand for food and lodging (Traldi, 2017). The workers related to fabrication and transportation of equipment are not hired locally where the towers will be implemented (Traldi, 2017).

Land use conflicts have been reported and the local impacts are real and diverse upon humans and other species (Santana, Silva, 2021). The implementation of renewable electric energy development projects is not necessarily free from negative externalities. A documentary recorded in the Northeast, in Pernambuco state, shows complaints regarding the noise, the wind, the shadow, the reduced soil humidity (deforestation, wind), the disorders in animal life due to the shadow, the skin problems caused by the radiation, the emotional problems and the sleepless nights (Comissão Pastoral da Terra et al., 2023).

THE CASE STUDY: BAHIA ELECTRIC ENERGY GRID

Located at in the Northeast region, the state of Bahia is one of the twenty-seven Federative Units of Brazil. It comprises an extent of 564,760,429 square kilometers, with a state population of 14,136,417 inhabitants, mostly living in urban zones, according to the last Census (2022). In relation to the size of its territory, Bahia is ranked the fifth state of Brazil. But, regarding population, the state occupies the position of fourth. The population density is 25.03 people per sq. km of land area (Censo, 2022).

In 2020, Bahia's Gross Domestic Product (GDP) represented 4.01% of Brazil's GDP (Bahia, 2022). At that same year, Bahia ranked in the 7th position (current prices) in comparation to the other Brazilian Federal Unities. Based upon the production perspective (value added), agriculture accounted for 10.4%, industry signified 22.2% and Services corresponded to 67.4% of the Bahian GDP. In the context of the industry, electricity accounted for 19.6% of the sector's GDP and participation has been growing in the last five years (Bahia, 2022).

The Bahian gross domestic supply of energy comprises production, import, inventory variation, export, untapped energy, and injections (Secretaria de Infraestrutura do Estado da Bahia, 2018). In that context, mostly a mixed wind-hydro-thermal power system produces electrical energy in Bahia, with a smaller contribution of solar energy (Empresa de Pesquisa Energética, 2023a, 2023d).[2] Hydropower is no longer the major origin of electricity in that state (BEN 2023d). The regional networks (south, southeast/central west, north and northeast) of electrical energy production and transmission lines are unified in a countrywide grid and shape the National Interconnected Electrical Energy System (*Sistema Interligado Nacional*)—SIN.

Hydropower

The history of large-scale hydropower generation in Bahia is similar to the Brazilian one. Between 1930 and early 1990s the national state provided, among other things, investments in basic inputs and infrastructure. The pro-growth Brazilian state developed the areas which the private sector did not want or could not undertake (Frieden, 1981). Some sectors were thought to be the Brazilian state's business. For example, the 1934 Constitution [Art. 118, Art. 119] and 1934 Water Code [Art. 43, Art. 63, Art. 139] established that industrial use of the waters such as for hydropower shall be undertaken/overseen by the federal state [1934 Constitution, 1934 Water Code] (Brasil, 1934).

Given the prevailing postwar understanding of development, the country required electric power for its progress. By the 1950s, electrical shortages and a growing demand for power were a challenge for Brazilian development (Brewer for *The New York Times*, 20 November 1954). The federal government chose the industrialization-urbanization path and influenced the use of the São Francisco River.

The stream heads north and then takes an eastern journey for a total of about 2,863 kilometers, finally flowing into the Atlantic Ocean. The first station producing electricity on a large-scale went on-line in 1954. Nine dams, not all of them in Bahia, and the human management of them govern the flow of that river. The São Francisco River system drains an area of approximately 640,000 square kilometers. Hydropower infrastructures and their management have led to changes in the landscape, and in both water and sediment flows. It has affected the provision of other ecosystem services too, disturbing fish, the fishery and fishermen and traditional riparian agriculture (Nascimento, 2010). The watershed comprises territories in six different states—Minas Gerais, Bahia, Goiás, Pernambuco, Alagoas, and Sergipe—and the Brazilian Federal District. Nearly 48% of the watershed lies in the state of *Bahia* and 36.9% in *Minas Gerais* (Comitê da Bacia Hidrográfica do São Francisco, 2004). The demand for electricity in a nation pursuing the path of development has driven this river system, major producer of hydropower in Bahia.

The Droughts

Drought is the lack of water or a water volume below the average available under normal local conditions for humans and other species (Tallaksen, Lanen 2004). In the Northeast of Brazil, it is a recurrent climatic event in the semi-arid region (Nascimento, 2019).

The Brazilian literature has linked drought in the Northeast of Brazil to temporal and spatial inter-annual variability of rainfall characterized by the occurrence of none, weak, or few shower events; precipitation converged in a short period of time, and rainfall set at large intervals. Droughts are influenced by the topography of the region, such as the Borborema Mountain Range, which impairs rain coming from the Atlantic coast; and high soil reflectivity, which precludes cloud formation (Nascimento n.d., Sociedade Brasileira para o Progresso da Ciência 2005, Serebrenick 1953).

Drought also has been connected to climatic anomalies such as *El-Ninõ*; distortion in the Atlantic Ocean surface temperatures, such as warmer in the north and colder in the south; and alterations in the position of air masses responsible for clouds and rain formation, the Intertropical Convergence Zone—ITCZ, when it is located further north than usual (Nascimento n.d., Davis 2001).

The consequences of droughts depend on their duration, volume, intensity, more specifically upon the period without the availability of water resources and also the extent of the region affected. Brazilian centralized electricity production system was not enough to prevent scarcity. In 2001, a drought caused water shortage, which led to black-outs and a temporary mandatory electricity consumption reduction program in Brazil (Tribuna do Norte 2001, Radiobras 2001).

Years of droughts in the Northeast of Brazil reduced the volume of water resources available in reservoirs and the supply of hydraulic energy and electricity during the years of 2001, 2008, 2013–2015. During the last period mentioned, the time of abnormal weather incidents of reduced precipitations, there was an increase in the use of nonrenewable energy, such as petroleum and oil products, for thermal generation, showing the vulnerability of the system to climate change and extremal events (Nascimento, 2019, 144).

In that South American nation, the use of primary renewable sources, such as wind energy, also jumped in the last decades. In conjunction, Bahia has expanded the generation of renewables. Wind energy, for example, started to show up in Bahia energy grid report in 2012 and the state's first wind farm was inaugurated in Brotas de Macaúbas (Secretaria de Infraestrutura do Estado da Bahia, 2016; Casa Civil da Bahia 2014).

Electricity Production in Bahia

At the beginning of April 2023, Bahian installed capacity (under operation + under construction + planned to be built) accounted for 1,032 enterprises corresponding to a Total Granted Power (Potência Outorgada Total) of 39,293,075.67 kW.[3] By implementation stage and all sources, the total of 16,118,373.07 kW (454 units) was under operation, the overall of 3,508,632.60 kW (97 units) was under construction and the amount of 19,666,070.00 kW (481 units) was planned to be built in the state (Figures 5.1–5.3).

Wind power installed capacity in operation in Bahia has surpassed hydropower in capacity and quantity. Wind was the major source (7,324,370.64 kW) and accounted for 45,58% of the Supervised Power (*Potência Fiscalizada*) (16,068,018.07 kW). Despite the efforts to increase solar energy, it is not the third most important source. The capacity to produce Thermal Power in operation is higher than the possibility of generating Photovoltaic Power. In addition, in Brazil, thermal electricity comes not necessarily from renewable sources.

Notes: In Brazilian acronym: CGH: Hydroelectric Plant (potentially less than or equal to 1,000 kW); EOL: Windeletric Power Plant; PCH: Small Hydroelectric Plant (more than 1,000 kW and less than or equal to 30,000 kW); UFV: Photovoltaic Power Plant; UHE: Hydroelectric Power Plant (capacity greater than 30,000 kW); UTE: Thermal Power Plant.

Among the four sources with facilities under construction, wind (EOL) and solar (UFV), in that order, present the highest installed capacity being built. Windelectric Power Plants accounts for about 72% of the projects under construction measured by the installed capacity. Among the four sources with facilities under construction, wind (EOL) and solar (UFV), in that order, present the highest installed capacity being built. Windeletric Power Plants

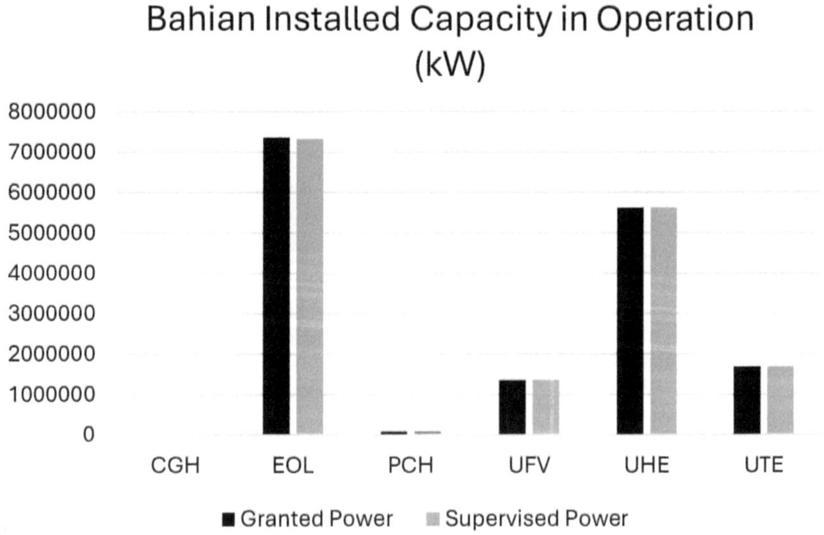

Figure 5.1. Bahian Installed Capacity in Operation (kW). Source: Modified from SCG—Superintendência de Concessões e Autorizações de Geração, Data accessed on 9 April 2023. Data for 8 April 2023.

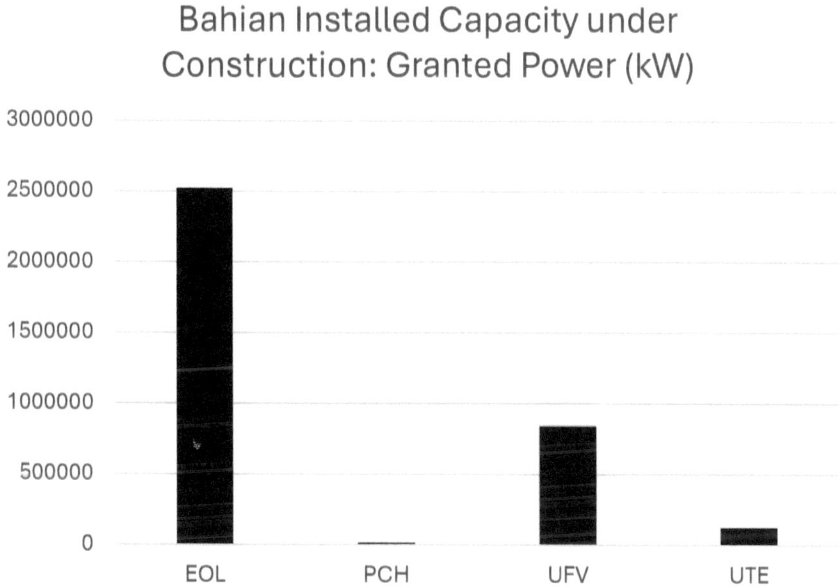

Figure 5.2. Bahian Installed Capacity Under Construction Granted Power (kW). Source: Modified from SCG—Superintendência de Concessões e Autorizações de Geração, Data accessed on 9 April 2023. Data for 8 April 2023.

account for about 72% of the projects under construction measured by the installed capacity.

Notes: In Brazilian acronym: EOL: Windeletric Power Plant; PCH: Small Hydroelectric Plant; UFV: Photovoltaic Power Plant; UTE: Thermal Power Plant.

According to the Brazilian Energy Research Office (EPE in its Portuguese acronym—*Empresa de Pesquisa Energética*) (2023a), regarding the installed capacity for electricity production in Brazil, in 2022, solar generation exhibited the most significant proportional expansion (59.5%), in comparison to 2021. This source presented the highest percentage increase in electricity production (79.8%), in the same period of time, in Brazil. In Bahia, solar power facilities (60%) surpass wind power (39%) among the ones planned to be built concerning installed capacity.

Notes: In Brazilian acronym: EOL: Windeletric Power Plant; UFV: Photovoltaic Power Plant; UTE: Thermal Power Plant.

SPATIAL DISTRIBUTION

The potential for wind and solar power generation have been identified in the state (Camargo Schubert Engenheiros Associados et al., 2013; AWS

Table 5.1. Wind Power Installed Capacity in Operation (kW) by Municipality—Spatial Distribution. Source: Modified from SCG—Superintendência de Concessões e Autorizações de Geração, Data accessed on 9 April 2023. Data for 8 April 2023.

Phase/Municipality	*Number of projects*	*Installed capacity (kW)*
In Operation	270	7,362,770.64
Bonito—BA	2	40,000.00
Brotas de Macaúbas—BA	3	95,190.00
Brumado—BA	2	60,000.00
Brumado—BA, Dom Basílio—BA	1	30,000.00
Caetité—BA	31	710,090.00
Cafarnaum—BA	3	89,900.00
Campo Formoso—BA	26	885,000.00
Casa Nova—BA	5	181,600.00
Gentio do Ouro—BA	16	403,500.00
Gentio do Ouro—BA, Xique-Xique—BA	1	36,000.00
Guanambi—BA	9	181,340.00
Ibipeba—BA	1	11,400.00
Igaporã—BA	20	377,120.00
Iraquara—BA	1	49,500.00
Iraquara—BA, Souto Soares—BA	1	33,000.00
Licínio de Almeida—BA	1	21,000.00
Morro do Chapéu—BA	33	1,161,650.00
Morro do Chapéu—BA, Várzea Nova—BA	1	72,000.00
Mulungu do Morro—BA	6	170,600.00
Mulungu do Morro—BA, Souto Soares—BA	1	49,500.00
Ourolândia—BA	2	54,600.00
Pindaí—BA	22	347,550.00
Riacho de Santana—BA	1	18,900.00
Sento Sé—BA	53	1,484,330.00
Sento Sé—BA, Umburanas—BA	5	29,700.64
Sobradinho—BA	4	84,000.00
Souto Soares—BA	1	49,500.00
Tucano—BA	5	241,800.00
Uibaí—BA	1	17,100.00
Umburanas—BA	2	58,800.00
Urandi—BA	1	18,000.00
Várzea Nova—BA	5	175,100.00
Xique-Xique—BA	4	125,000.00

Table 5.2. Solar Power—Spatial Distribution. Source: Modified from SCG—Superintendência de Concessões e Autorizações de Geração, Data accessed on 9 April 2023. Data for 8 April 2023.

Phase/Municipality	*Number of projects*	*Installed capacity (kW)*
In Operation	47	1,359,469.75
Barreiras—BA	4	94,640.00
Bom Jesus da Lapa—BA	8	214,000.00
Caravelas—BA	1	400.00
Casa Nova—BA	1	30,000.00
Guanambi—BA	1	14,300.00
Itaguaçu da Bahia—BA	1	30,520.00
Juazeiro—BA	9	279,180.00
Oliveira dos Brejinhos—BA	8	414,999.95
Salvador—BA	2	3,704.80
Tabocas do Brejo Velho—BA	10	275,800.00
Terra Nova—BA	1	225.00
Vitória da Conquista—BA	1	1,700.00

Truepower et al., 2018). An analysis of the spatial distribution of the existing wind and solar power structures shows a more consolidated presence of wind power generation (Table 5.1.). The solar generation is of a more incipient nature (Table 5.2.).

Bahian Policy Instrument

In addition to federal policies and programs, Bahian government actions have undertaken a major step which aided the growth of the alternative energy sector. In December 2002, Bahia State approved Law No. 7,980 (Bahia, 2001). The legal instrument created the Programa de Desenvolvimento Industrial e de Integração Econômica do Estado da Bahia (Industrial Development and Economic Integration Program of the State of Bahia)—DESENVOLVE. The objective of the program was to promote and diversify the industrial and agro-industrial matrix, with the formation of industrial densities in the economic regions and integration of the productive chains essential to the economic and social development and to the generation of employment and income in the State (Bahia, 2001). Decree No 8,205 (2002) regulates the DESENVOLVE program (Bahia, 2002).

The incentives intended to encourage the installation of new industries and the expansion, reactivation or modernization of industrial undertakings already installed, with the generation of new products or processes, improvement of technological characteristics and reduction of costs of existing products or processes (Bahia, 2001). The ones applying for benefits have to meet the conditions specified by law, such as job creation, technological development, social and environmental responsibility (Bahia, 2001).

The DESENVOLVE introduced means for tax exemption and reduction and extension of the tax's payment period. In practice, it benefited the implementation of power stations and the construction of transmission lines (Bahia, 2001). According to Lima (2018), besides the financial incentives, similar to the federal ones, Bahia state has provided other types of support to the energy industry, such as institutional support.

CONCLUSIONS

In recent years, Bahia has enlarged the installed capacity of generation of renewable electricity beyond hydropower, with emphasis on wind and solar power. Wind energy has overcome the view of being only a complementary source. It is now the major one in the state. But it is important not to concentrate on the existing sources. The state already seeks the installation of a green hydrogen plant.

In summary, the federal and state government provided incentives for those in the renewable energy marketplace. But it is important to deepen the knowledge of and address the negative externalities, and better understand the multiple aspects of local's socio-economic dynamic of the solar/wind power generation.

NOTES

1. The main Greenhouse gases (GHG) are carbon dioxide—CO_2, methane—CH4, nitrous oxide—N2O and chlorofluorocarbons—CFCs. Water vapor is a natural greenhouse gas.
2. Thermal energy means the energy obtained by the combustion of fossil fuel or biomass; or released in nuclear reactions.
3. Data from Superintendência de Concessões e Autorizações de Geração (SGC) extracted from Power BI on 8 and 9 April. The data refer to 8 April 2023.

BIBLIOGRAPHY

Agencia Brasil—Radiobras (2001) Sinopse—Resumo Dos Jornais. Sinopse Radiobras.

AWS Truepower et al. (2018) Atlas solar: Bahia. Salvador: SECTI, http://sit.infraestrutura.ba.gov.br/docs/download/mapas/atlas-solar-Bahia-2018.pdf, accessed 17 May 2023.

Bahia, Casa Civil da (2014) Balanço das Ações 2007–2014, http://hostnave.com.br/pdf-ipad-balanco-8-anos/Balanc%CC%A7o-8anos-iPad.pdf, accessed 30 April 2018.

Bahia, Governo do Estado da. Law N° 7.980 de 12 de dezembro de 2001. Institui o Programa de Desenvolvimento Industrial e de Integração Econômica do Estado da Bahia—DESENVOLVE . . . , http://www.legislabahia.ba.gov.br/documentos/lei-no-7980-de-12-de-dezembro-de-2001#, accessed 27 August 2023.

Bahia, Governo do Estado da. Decree N° 8.205 de 03 de abil de 2002. Aprova o Regulamento do Programa de Desenvolvimento Industrial e de Integração Econômica do Estado da Bahia—DESENVOLVE e constitui o seu Conselho Deliberativo, http://www.legislabahia.ba.gov.br/documentos/decreto-no-8205-de-03-de-abril-de-2002#, accessed 27 August 2023.

Bahia, Superintendência de Estudos Econômicos e Sociais da (2022) Boletim PIB Estadual Annual, https://sei.ba.gov.br/images/pib/pdf/estadual/anual/economia_baiana_2020.pdf, accessed 21 May 2023.

Brasil. República Federativa do. Código de Águas, 24,643 of 1934.

Brasil. República Federativa do. Constituição da República Federativa do Brasil 1934.

Brasil. República Federativa do. Law No. 10,438 of April 26, 2002. Dispõe sobre a Expansão da Oferta de Energia Elétrica . . . Diário Oficial da União edição extra (2002).

Brasil, República Federativa do. Law N° 10.848, DE 15 DE MARÇO DE 2004. Dispõe sobre a comercialização de energia elétrica . . . , https://www.planalto.gov.br/ccivil_03/_ato2004-2006/2004/lei/L10.848compilado.htm, accessed 28 August 2023.

Brasil. República Federativa do. Decree N° 5.163, of July 30, 2004. Regulamenta a comercialização de energia elétrica, o processo de outorga de concessões e de autorizações de geração de energia elétrica, e dá outras providências . . . , http://www.planalto.gov.br/ccivil_03/_ato2004-2006/2004/decreto/d5163.HTM, accessed 6 July 2023.

Brewer, S. P. (1954) Power shortage eased in Brazil. *The New York Times*, 20 November, 8.

CAMARGO-SCHUBERT ENGENHEIROS ASSOCIADOS . . . [et al.]. (2013) Atlas eólico: Bahia. Dados do modelo mesoescala fornecidos por AWS Truepower. Salvador: SECTI. http://www.infraestrutura.ba.gov.br/arquivos/File/publicacoes/atlaseolicobahia2013.pdf, accessed 17 May 2023.

Comitê da Bacia Hidrográfica do Rio São Francisco (2004) Plano De Recursos Hídricos Da Bacia Hidrográfica Do Rio São Francisco. Salvador: CBHSF.

Censo 2022—Instituto Brasileiro de Geografia e Estatística (2023) Estado@Bahia, https://cidades.ibge.gov.br/brasil/ba/panorama, accessed 30 August 2023.

Clark, T. W. (2002) *The Policy Process: A Practical Guide for Natural Resources Professionals*. New Haven, CT: Yale University Press.

Comissão Pastoral da Terra; Instituto Mãe Terra; Universidade de Pernambuco; Fundo Casa Socioambiental (2023) Vento agreste. Pernambuco: CPT, https://www.youtube.com/watch?v=pYEDygsfheE, accessed 21 May 2023.

Davis, M. (2001) *Late Victorian Holocausts: El Niño Famines and the Making of the Third World*. London: Verso.

Dutra, R. M.; Szklo, A. S. (2008) Incentive Policies for Promoting Wind Power Production in Brazil: Scenarios for the Alternative Energy Sources Incentive Program (PROINFA) Under the New Brazilian Electric Power Sector Regulation. Renewable Energy 33 (1): 65–76.

Empresa de Pesquisa Energética—EPE (2014). Leilão de Energia de Reserva de 2014 Participação dos Empreendimentos Solares Fotovoltaicos: Visão Geral, https://www.epe.gov.br/sites-pt/publicacoes-dados-abertos/publicacoes/PublicacoesArquivos/publicacao-94/NT_EPE-DEE-NT-150_2014.pdf, accessed 25 July 2023.

Empresa de Pesquisa Energética—EPE (2023a) Statistical Yearbook of Electricity—2022 baseline year. Rio de Janeiro: EPE, https://dashboard.epe.gov.br/apps/anuario-livro-eng/, accessed 19 June 2023.

Empresa de Pesquisa Energética—EPE (2023b) Anuário Estatístico de Energia Elétrica 2023—Workbook, https://www.epe.gov.br/pt/publicacoes-dados-abertos/publicacoes/anuario-estatistico-de-energia-eletrica, accessed 19 Jun 19 2023.

Empresa de Pesquisa Energética—EPE (2023c). Fact Sheet: Anuário Estatístico de Energia Elétrica, https://www.epe.gov.br/sites-pt/publicacoes-dados-abertos/publicacoes/PublicacoesArquivos/publicacao-160/topico-168/Fact%20Sheet%202023%20-%20Anu%C3%A1rio%20Estat%C3%ADstico%20de%20Energia%20El%C3%A9trica.pdf, accessed 19 June 2023.

Empresa de Pesquisa Energética—EPE (2023b). BEN—Séries históricas e matrizes: dados estaduais, https://www.epe.gov.br/pt/publicacoes-dados-abertos/publicacoes/BEN-Series-Historicas-Completas, accessed 30 August 2023.

Frieden, J. (1981) "Third World Indebted Industrialization: International Finance and State Capitalism in Mexico, Brazil, Algeria, and South Korea." *International Organization* 35 (3): 407–431.

Gleick, P. H. (2009) *The World's Water 2008–2009: The Biennial Report on Freshwater Resources*. Washington, DC: Island Press.

Hao, D. et al. (2022) "Solar Energy Harvesting Technologies for PV Self-Powered Applications: A Comprehensive Review." *Renewable Energy* 188: 678–697, https://doi.org/10.1016/j.renene.2022.02.066, accessed 23 July 2023.

Lasswell, H. D. (1965) "The Policy Sciences of Development." *World Politics* 17, no. 2: 286–309.

Lasswell, H. D. (1970) "The Emerging Conception of the Policy Sciences." *Policy Sciences* 1(1): 3–14.

Lasswell, H. D. (1971) *A Pre-View of Policy Sciences*. Edited by Yehezkel Dror, Policy Sciences Book Series. New York: American Elsevier Publishing Company, Inc.

Lima, R. C. (2018) "A indústria de aerogeradores e o desenvolvimento regional: perspectivas de consolidação na Bahia." Master Thesis, Universidade Federal da Bahia, https://repositorio.ufba.br/bitstream/ri/26300/1/Disserta%c3%a7%c3%a3o%20Vers%c3%a3o%20Final%20-%20Rafaela%20Costa%20Lima.pdf, accessed 27 August 2023.

Maxwell, J. A. (1996) *Qualitative Research Design: An Interactive Approach*. London: Sage Publications.

Mazzucato, M. (2018) "Mission-Oriented Innovation Policies: Challenges and Opportunities." *Industrial and Corporate Change* 27(5): 803–815, https://edisciplinas.usp.br/pluginfile.php/4667792/mod_resource/content/1/Mazzucatto.pdf, accessed 23 July 2023.

McKenna, R. et al. (2022) "High-Resolution Large-Scale Onshore Wind Energy Assessments: A Review of Potential Definitions, Methodologies and Future Research Needs." *Renewable Energy* 182: 659–684, https://www.sciencedirect.com/science/article/pii/S0960148121014841, accessed 25 August 2023.

Ministério de Minas e Energia do Brasil (2010). *Plano Decenal de Expansão de Energia 2019*. Brasília: Ministério de Minas e Energia.

Nascimento, L. N. (n.d.). "Drought." In *Brazil Today: An Encyclopedia of Life in the Republic*, edited by John Crocitti. Westport, CT: Greenwood Publishing Group, Inc.

Nascimento, L.N. (2010). "The Long Journey to Become the 'River of National Unity': The São Francisco River Basin from 1940 to 2008 and the Interactions of Environment, Government and Local Citizens." PhD diss., University of New Hampshire.

Nascimento, L. N. (2019). "Wind Energy and Policy in Brazil: An Assessment of the State of Bahia." In *Renewable Energy: International Perspective on Sustainability*, ed. D. Kurochkin, E. V. Shabliy, E. Shittu, pp. 131–155. New York: Springer International Publishing.

Organization for Economic Cooperation and Development OECD (2023) "Demand-Side Innovation Policies." https://doi.org/10.1787/9789264098886-en, accessed 23 Jun.

Overland, I.; Juraev, J.; Vakulchuk, R. (2022) "Are Renewable Energy Sources More Evenly Distributed Than Fossil Fuels?" *Renewable Energy* 200: 379–386, https://doi.org/10.1016/j.renene.2022.09.046.

Pesquisa Nacional por Amostra de Domicílios Contínua (2019) Rio de Janeiro: IBGE, ftp.ibge.gov.br/Trabalho_e_Rendimento/Pesquisa_Nacional_por_Amostra_de_Domicilios_continua/Trimestral/, accessed 8 January 2020.

Pinheiro, A. de F.; Ararlba, J. V. D. (2017) "Panorama da energia eólica na Bahia: uma radiografia de quatro anos (2012–2016)." *Bahia Análise & Dados Energias Renováveis* 27(1): pp. 356–375.

REN21 (2023) "What Are the Current Trends in Renewable Energy?" https://www.ren21.net/what-are-the-current-trends-in-renewable-energy/#re-mainstream-in-power, accessed 15 July.

Santana, A. O. de; Silva, T. A. A. da. (2021). "Produção de energia eólica em Pernambuco e a injustiça ambiental sobre comunidades rurais." *Rev. katálysis* 24(1), https://doi.org/10.1590/1982-0259.2021.e73663, accessed 31 August 2023.

Santos, J. A. F. de A.; Torres, E. A. (2017) "Potencial e inserção de novas energias renováveis na matriz elétrica baiana para geração elétrica centralizada até 2050." *Bahia Análise & Dados Energias Renováveis* 27(1): 144–173.

Sociedade Brasileira Para o Progresso da Ciência (2005). "As Incertezas Do Projeto De Transposição." Ciência Hoje—Revista de Divulgação Científica da SBPC: O Rumo do Velho Chico: Benefícios e Incertezas da Transposição das Águas do Rio São Francisco, Julho, 48–52.

Secretaria de Infraestrutura do Estado da Bahia, Balanço Energético da Bahia 2016 (2018) http://www.infraestrutura.ba.gov.br/arquivos/File/publicacoes/beeba2016.zip, accessed 11 February.

Serebrenick, Salomão (1953). *Condições Climáticas Do Vale Do São Francisco: Clima—Enchentes E Estiagens—Reflorestamento.* Rio de Janeiro: Comissão do Vale do São Francisco—CVSF—Departamento de Imprensa Nacional.

Shimomaebara, L. A.; Peyerl, D. (2021) "Energia Solar no Brasil: Histórico e planejamento energético." *Revista de Políticas Públicas* 25(2): 854–869, https://www.redalyc.org/journal/3211/321171229021/html/, accessed 25 July 2023.

Staffell, I.; Green, R. (2014) "How Does Wind Farm Performance Decline with Age?" *Renewable Energy* 66: 775–786. https://www.sciencedirect.com/science/article/pii/S0960148113005727, accessed 23 July 2023.

Tallaksen, L. M.; Henny A. J. V. L. (eds.). (2004) *Hydrological Drought: Processes and Estimation Methods for Streamflow and Groundwater (Developments in Water Science)*. Amsterdam, the Netherlands: Elsevier.

Traldi, M. (2017) "Implantação de parques eólicos no semiárido brasileiro e a promessa da geração de empregos." *Bahia Análise & Dados Energias Renováveis* 27(1): 174–202.

Tribuna do Norte (2001) Chesf Ameaça Recorrer a Apagões. 4 October.

United Nations (1948) The Universal Declaration of Human Rights 1948, https://www.ohchr.org/EN/UDHR/Documents/UDHR_Translations/eng.pdf, accessed 21 April 2020.

United Nations (2019) Sustainable Development Goals: Knowledge Platform, https://sustainabledevelopment.un.org/sdg7, accessed 21 April 2020.

United Nations (2023). What Is Renewable Energy? https://www.un.org/en/climatechange/what-is-renewable-energy, accessed 14 July 2023.

Wishlade, F.; Michie, R.; Verbon, P. (2017) Research for REGI Committee—Financial Instruments for Energy Efficiency and Renewable Energy, European Parliament, Policy Department for Structural and Cohesion Policies, Brussels, https://www.europarl.europa.eu/RegData/etudes/STUD/2017/601992/IPOL_STU(2017)601992_EN.pdf, accessed 25 June 2023.

World Meteorological Organization. (2023) State of the Global Climate 2022, WMO-No. 1316. Geneva, Switzerland: WMO, https://library.wmo.int/doc_num.php?explnum_id=11593,accessed 22 April 2023.

World Meteorological Organization. (2022) State of Global Water Resources 2021, WMO-No. 1308. Geneva, Switzerland: WMO, https://library.wmo.int/doc_num.php?explnum_id=11586, accessed 23 Apr 2023.

Zhou, D. et al. (2021) "Literature Review on Renewable Energy Development and China's Roadmap." *Front. Eng*, 8(2): 212–222, https://doi.org/10.1007/s42524-020-0146-9, accessed 15 May 2023.

Chapter Six

The Path to Balanced Decentralization of Energy Systems

Envisioning Interoperable Peer-to-Peer Energy Markets

Sakshi Mishra, Roohallah Khatami, and Yu Christine Chen

Energy infrastructure is crucial to a nation's critical systems. Electricity is essential not only for keeping lights on for consumers but also as a key component of industrial processes. In the current digital era, where electronic devices are fundamental to everyday life, the need for a stable electricity supply—to power these devices—has significantly increased. Concurrently, energy systems are undergoing substantial changes. The future of decentralized energy systems is poised at a transformative juncture, propelled by an array of dynamic forces. These range from technological advancements to regulatory shifts, economic factors, and a growing societal commitment to sustainability. As we look ahead, it is clear that these systems are not just evolving in isolation but are being reshaped by a confluence of influential trends. Central to this evolution is the relentless pace of technological innovation. Breakthroughs in renewable energy technologies, enhanced energy storage solutions, and smart grid technologies are redefining the landscape of energy distribution and management. These trends in the global energy landscape are due to various factors (detailed in Appendix I).

Historically, the electricity sector has been monopolistic, dominated by vertically integrated utilities. However, the liberalization of electricity markets in recent decades has allowed only utility-scale generation plant owners to participate. From a consumer standpoint, the electricity sector is still predominantly centralized: consumers have minimal (if any) influence on market dynamics, leaving them as mere "price-takers." The rise of Distributed Energy Resources (DERs), achieving high levels of penetration, provides a glimmer of hope: it enables consumers and communities to move towards self-sufficiency in unprecedented ways, making the option of going "off-grid" increasingly feasible. Depending on the region, prosumers might already be able to enter agreements with DER aggregators—third-party for-profit organizations that act as intermediaries between prosumers and the wholesale markets. These agreements typically bind prosumers to sell their electricity for a set period (usually a few years), which may not maximize the value of the energy provided by prosumers. Thus, the structure of these agreements does not fully support the goal of decentralized operations where prosumers are the main decision-makers.

In this context, we re-envision the operational model of decentralized energy systems, adopting a bottom-up approach to grid management. Furthermore, in this study, we propose that the centralized (top-down) and decentralized (bottom-up) models for grid operations are not inherently contradictory. We propose that essential to a paradigm shift in the electricity sector's operating philosophy is to find synergies between top-down and bottom-up approaches to managing the generation, transmission, distribution, and end-use pipeline.

In practical terms, this proposal means that prosumers or end users are no longer merely passive recipients of set prices. Instead, they engage in the wholesale markets as active price setters. At first, the notion of top-down and bottom-up approaches existing simultaneously might appear contradictory, and one might be inclined to describe the bottom-up method of enhancing grid edge control with DER deployments as "disruptive." However, it's important to recognize that technological disruptions in the electricity sector are unlikely to occur rapidly due to the industry's conservative nature and vast scale. Moreover, the last decade has seen significant (Statisca, 2023; Justia US Law, 2020; Bade, 2018) resistance from utilities committed to the top-down architecture against DERs and community microgrids, as these concepts challenge their established business models. If the bottom-up approach aims to complement and enhance the goals of the existing top-down system, rather than completely replacing the centralized system, it is more likely to be broadly adopted (Morstyn et al., 2018). Nevertheless, the integration of top-down and bottom-up architectures will necessitate innovative coordination mechanisms capable of aggregating small-scale DERs for broader system objectives while fully respecting the energy preferences of their owners (i.e., their choices about when and how to use the energy from their DERs).

At the same time, it's crucial to identify various revenue streams to motivate end-users (i.e., prosumers) to keep investing in DER projects. Theoretically, a simple method for creating value streams is to link these prosumers to the existing wholesale markets for energy, transmission rights, and ancillary services. Yet, this approach is limited in practical effectiveness due to several obstacles that hinder DER participation. "Individual consumers and small-scale generators have minimal impact at the transmission level, and the complexity of the required processing and communications infrastructure for market participation would lead to transaction costs that surpass the potential benefits. Additionally, market price volatility poses risks that are challenging to mitigate at the individual level. Traditionally, these consumers have been catered to in retail markets by large suppliers, where economies of scale make it feasible to absorb transaction costs and diversify market price risks" (Matheson, 2019). This difficulty in accessing wholesale markets leaves prosumers with the sole advantage of using their flexibility to shift their peak demand, thereby reducing their demand charges.

Coordinated local DERs can successfully address some of these challenges by providing mutually beneficial solutions for both grid operators and end-users or prosumers. First, by functioning as a coordinated group of generation and storage technologies, DERs can engage in the wholesale markets, creating a revenue stream for their individual owners. Through their collective participation, they offer grid operators the chance to utilize their services (such as in the ancillary market) without the need to manage and control each small-scale generation technology[1] individually.

Secondly, coordinated local DERs can diminish the need for upstream generation and transmission capacity by fulfilling a significant portion of the demand with local generation within the coordinated cluster. It's important to note that capacity expansion by utilities aligns well with their current business models. However, there are several systemic challenges associated with developing new grid infrastructure: i) Siting new transmission lines involves securing approval for new routes and acquiring the necessary land rights; ii) Finding a fair method to distribute the construction costs of a new transmission line built in one state when it benefits consumers in other states; iii) Extending the network of long-distance transmission lines to renewable energy generation locations, which often have high-quality wind and solar resources but are typically distant from high-demand areas; and so forth. Thus, reducing upstream capacity needs is seen as a positive outcome from both the regulator's and the prosumer's viewpoints.

Thirdly, coordinated DERs can facilitate cost reductions on utility bills (through peak demand reduction) and offer resilience benefits, such as maintaining critical loads during utility outages. These cost savings occur when the peak demand of an individual prosumer is supplied by local generation from the coordinated fleet, thus lowering the utility demand charges applied to each metered prosumer. These savings also reflect the energy efficiency achieved by the grid, as it avoids transporting power over long distances and the need for Peaker plants to meet consumer demand during peak hours. Additionally, these DERs can enhance the resilience of the distribution grid by providing black-start capabilities in the event of a blackout.

Considering the numerous advantages of managing DERs as a coordinated fleet, this idea has recently gained popularity and is known as Virtual Power Plants (VPPs). A VPP represents a group of DERs that are orchestrated to achieve visibility, controllability, and influence at the transmission level of the power network (Morstyn et al., 2018).

Alternatively, individual prosumers have another method to be compensated for the surplus energy produced by their DERs, known as the "net-metering" policy. In this system, prosumers effectively sell their electricity back to the grid at a "buy-back rate." However, the cost at which they purchase electricity from the grid is usually much higher than the buy-back rates they receive for selling electricity back to the grid. Consequently, a new form of revenue generation for prosumers has been emerging, known as peer-to-peer (P2P) energy markets.

P2P energy markets enable prosumers to trade electricity directly with each other without intermediaries, at prices they mutually agreed upon, by providing a platform for these transactions. This option is increasingly feasible with higher DER penetration and promotes decentralization by giving prosumers more control over the energy produced by their DERs. P2P energy markets are thus crucial in advancing the bottom-up approach in energy system operations. When designed effectively, P2P energy markets not only empower prosumers but also offer several benefits to the central grid, such as helping with congestion management and providing ancillary services.

While P2P energy markets can operate independently, they are primarily expected to function among prosumers connected to the utility grid. In this grid-connected scenario, prosumers can also benefit by contributing their DER generation to centralized markets—hence the coordinated fleets (as discussed in the context of VPPs above). Therefore, it is essential to recognize that integrating P2P energy markets with VPPs is an optimal strategy for maximizing revenue streams for prosumers, further fostering the decentralization paradigm.

PROBLEM DEFINITION

For over a century, the grid has functioned in a centralized, top-down manner. However, with the increasing integration of distributed energy resources (DERs), the grid edge is becoming more enriched with smart computing and communication technologies. Consequently, a bottom-up approach, favoring the decentralization of energy systems, is likely to rise in prominence alongside the traditional centralized model. In recent academic studies, there has been an increasing focus on local decentralized energy trading (Pudjianto et al. 2007; Doan et al. 2021; Wang et al. 2019; Seven et al. 2020; AlAshery et al. 2020; Thomas et al. 2019; Han et al. 2020), and there are ongoing testbed and prototype deployments globally to test the viability of localized peer-to-peer energy trading (Mengelkamp et al. 2018) and government-led microgrid projects (Morstyn and McCulloch 2019). This shift towards a bottom-up approach poses challenges to the established top-down grid operation structure. As decentralization progresses, conflicts are expected among various stakeholders in the electricity supply chain and markets, such as generation owners, large utilities, small DER owners, and policymakers. Therefore, it's crucial to harmonize the coexistence of these two seemingly opposing grid management approaches, aiming to transform the future grid into a reliable, efficient, decarbonized, and cost-effective infrastructure to meet increasing global energy demands.

Given the rapid changes in the energy sector, research is advancing across various domains, from generation to transmission, distribution, and end-user engagement. Innovations in these areas, addressing technical, economic, and policy dimensions of energy systems, are essential to achieving decarbonization and sustainability goals (Cali et al., 2019). However, there's an urgent need to integrate advancements across these specialized fields, understanding their collective impact and shaping decentralized energy system operations that complement existing top-down processes. This integration will enhance the functionality of standalone smart grid technologies, contributing to a cleaner, equitable energy future through a systems approach feedback loop. This process makes the research in these niche areas increasingly applicable, potentially accelerating the research-to-deployment timeline for these innovative concepts and ideas. It's also crucial to synergize these smart grid technologies and operational paradigms to achieve broader objectives like reducing carbon emissions, ensuring reliability, and improving the overall energy efficiency of the electricity supply chain.

Therefore, this work aims to develop a holistic, prosumer-centric grid operations framework. This framework will navigate the complex dynamics of power systems markets and operations, bridging the gap between top-down and bottom-up approaches in electricity grid management at the distribution level. It not only empowers prosumers with more control over the energy from their DERs but also aligns well with existing large-scale centralized markets.

CONTRIBUTIONS OF THIS WORK

Literature on P2P energy trading often overlooks the combination of P2P trading with decentralized Virtual Power Plants (VPPs) (Andoni et al., 2019; Directorate-General, 2015). Additionally, the P2P trading models proposed in these studies do not take into account the real-world regulatory frameworks that govern or restrict retail customer interactions with wholesale markets (FERC, 2020; Ton et al., 2018).

To achieve the goal of developing a prosumer-centric P2P energy trading framework that operates in conjunction with the top-down grid management approach, we have identified and achieved the following specific objectives:

- Define the value proposition of decentralized energy systems and introduce advanced concepts of operational planning within the decentralization context.
- Develop a prototype for the hierarchical structure of decentralized Peer-to-Peer markets, recognizing the predetermined interactions between top-down and bottom-up approaches in electricity grid operations.
- Construct a model for the hierarchy of decentralized dispatch decisions in P2P markets, taking into account the intricate interplay among the physical, market, and cyber spheres.

The resistance from incumbent utilities and traditional grid operational mechanisms within the regulatory environment has largely prevented fair participation of prosumers. Our framework fundamentally enables the coexistence of centralized and decentralized electricity market models, creating a conducive environment for the burgeoning decentralization paradigm to flourish and contribute to a sustainable, low-carbon future. Notably, our proposed framework emphasizes the "decentralization paradigm," empowering prosumers to sell their energy and maximize savings—a key feature of a truly decentralized energy market. Simultaneously, our framework does not seek to abolish top-down markets; rather, it innovatively connects both through multi-level operational structures. The framework presented in this chapter was originally proposed in (Mishra et al., 2023), and interested readers may find an in-depth technical discussion with modelling details therein.

BUILDING BLOCKS

The Interface of Established Centralized Energy Markets and New Peer-to-Peer Exchanges

The full potential of decentralized and peer-to-peer energy markets, particularly for prosumers on the distribution grid, can be fully harnessed only when considered in relation to their interaction with the central grid's market operations. This necessity arises as the operations of the transmission and distribution grids become increasingly interconnected due to the bidirectional flow of power. Additionally, the coordinated fleet of DERs provides benefits like demand response and load flexibility to grid operators, while simultaneously creating an additional revenue stream for prosumers. This mutually beneficial scenario underscores the importance of considering the interactions between centralized (wholesale and retail) markets and locally coordinated DER operations (i.e., VPPs). The integration of decentralized systems into the established grid requires not only technological compatibility but also regulatory and market-based alignments to ensure seamless operation and value maximization.

Centralized electricity markets typically comprise two segments—retail and wholesale. Retail markets deal with selling electricity to end consumers, whereas wholesale markets involve the trade of electricity among electric utilities and energy traders before it reaches consumers through the retail markets. This traditional framework is now evolving to accommodate distributed generation sources, which add complexity to market operations and pricing structures. The emergence of prosumers, who both consume and produce electricity, is challenging the traditional dynamics of these markets and necessitating a reevaluation of market mechanisms and regulations. The operations of these wholesale and retail markets involve numerous participants in the power systems. These participants include utilities, grid operators, load-serving entities, third-party service providers (in retail markets), and energy traders, among others. The market mechanism that delivers electricity to end users is governed by a set of complex interactions among these agents. These agents include electric utilities, vertically integrated utilities, load-serving entities (or local distribution companies), electricity retailers, grid operators or transmission system operators (TSOs), and electric utilities as grid Operators. A detailed description of these agents and their roles is provided in Appendix II.

For a P2P energy market to function effectively in practice, its design must account for the dynamics of agent interactions in the electricity supply chain. This includes understanding the regulatory frameworks, market structures, and economic incentives that influence these interactions. Furthermore, technological advancements in grid infrastructure and data management systems are essential for the efficient operation of P2P markets within this complex ecosystem. Therefore, the subsequent subsections are dedicated to discussing the concepts related to wholesale and retail markets and their connection to physical grid operations, which form the basis for designing the proposed decentralized peer-to-peer energy markets framework.

THE UNFOLDING OF RETAIL MARKET LIBERALIZATION

In areas with retail deregulation, electricity consumers have the choice to select their electric supplier (referred to as customer choice), instead of being mandated to buy electricity from their local electric utility. This introduces competition in retail electricity pricing. In regions offering customer choice, multiple electric suppliers may exist, prompting these retailers to offer competitive prices to attract customers (contracts with generation suppliers often provide a fixed charge in dollars per kilowatt-hour for a specified duration).

For consumers, choosing a supplier other than their local utility has its advantages and disadvantages. Retail competition can lead to lower electricity bills and allow customers to choose their energy source, such as opting for a clean energy supplier. However, *independent suppliers* typically require customers to enter into contracts, which might fix their electricity price for several years. While fixed rates can be advantageous for some, they might be less favorable for others if the agreed rate turns out to be higher than the local utility's rate. It is also crucial to recognize that customer choice applies only to the generation part of the electricity bill. Transmission and distribution services are still monopolized by the local utility, as these are considered natural monopolies. Therefore, only a segment of the electric rates in these areas is subject to competitive pricing. For customers who do not choose an independent power supplier, their local utility remains responsible for supplying electricity, which it acquires from generation companies. Our framework aims to replace the independent for-profit aggregator companies discussed earlier with the functioning of VPPs. We design the framework to address the challenge of making the VPPs operate semi-autonomous digital entity, not as a centralized aggregator seeking profit, but as a community-owned digital system that maximizes prosumer savings. The operations of VPPs are architected accordingly.

Over time, these Virtual Power Plants (VPPs) should have the capability to interact with both retail and wholesale markets. However, in our research for this project, we have chosen to concentrate on the framework's interaction with wholesale markets. This decision is primarily due to the more widespread presence of wholesale markets in comparison to retail markets within the North American power grid. Another reason for focusing on the interactions of our proposed framework with the wholesale market is our objective to maximize the value that prosumers can derive from the energy produced by their Distributed Energy Resources (DERs). Wholesale markets, particularly those operating on short time frames, tend to offer more lucrative opportunities.

STRUCTURAL OVERVIEW OF WHOLESALE MARKETS

In deregulated areas, grid operators, known as Regional Transmission Organizations (RTOs[2]) and Independent System Operators (ISOs[3]), generally operate three types of markets that set the wholesale prices for electricity sales: i) Energy Markets; ii) Capacity Markets; and iii) Ancillary Markets. An overview of these three market types is included in Appendix III.

DECENTRALIZED PEER-TO-PEER ENERGY MARKETS: EXPLORING THE TECHNOLOGICAL LANDSCAPE

The hierarchical framework outlined and described in section 3 effectively utilizes current energy technologies (such as Virtual Power Plants [VPPs], microgrids, and peer-to-peer [P2P] energy trading) to fulfill the goal of creating scalable and interoperable decentralized P2P energy markets at the distribution grid level. This approach is intended to enhance the autonomy, efficiency, and resilience of the local generation. It is important to note that there are some variations in how these technologies are interpreted and understood in the literature. Therefore, in the subsequent subsections, we will clarify the fundamental functionalities and capabilities of these technical tools, as they are crucial components of the proposed framework.

DERS

Distributed Energy Resources (DERs) are small to medium-sized resources directly connected to the distribution network (Dimeas, 2005). In the United States, the Federal Energy Regulatory Commission (FERC) Order No. 2222 broadly defines DERs as "any resource located on the distributed system, any subsystem thereof or behind a customer meter" (Wolf, 2020). DERs encompass electric storage resources, distributed generation, thermal storage, heat pumps, flexible and controllable loads, energy efficiency measures, and electric vehicles along with their charging equipment. Typically, these resources are situated near the electricity end users. Demand Response is also considered a type of DER.

Net energy metering (NEM) is a system where DERs are connected behind-the-meter to a distribution system. In this setup, any excess power generated by prosumer-owned DERs is fed back into the grid, allowing customers to offset the cost of power drawn from the distribution utility. This surplus flow usually happens when the DERs' production exceeds the customer's total demand. Due to NEM, which is a policy highly favorable to DERs, there has been a significant increase in DER installations in some US states.

The recent FERC Order No. 2222 (FERC, 2020) has fortuitously allowed DERs to participate in wholesale markets. This ruling enables DERs to join traditional resources in regionally organized wholesale markets through aggregations. It opens US-organized wholesale markets to new energy sources and grid services, allowing various distributed electricity sources to aggregate to meet minimum size and performance requirements that might be unattainable individually. The critical aspect here is aggregation. This development creates opportunities for third-party service providers, distinct from Load Serving Entities responsible for serving retail customers. The mechanisms of aggregation, which enhance scalability and interoperability, are a focal point in our proposed framework. We will delve into this topic more thoroughly in the following sections.

MICROGRIDS

The concept of a microgrid is continuously evolving with the introduction of new technologies and ideas in the smart grid and urban planning sectors. According to the US Department of Energy, a microgrid is defined as "a group of interconnected loads and distributed energy resources within clearly defined electrical boundaries that acts as a single controllable entity concerning the grid. A microgrid can connect and disconnect from the grid to enable it to operate in both grid-connected or island-mode" (Morstyn et al., 2019). In Europe, a microgrid is seen as "a low voltage distribution network consisting of various DG [Distributed Generators], storage devices, and controllable loads that can operate interconnected or isolated from the main distribution grids" (United Nations Development Programme and Government of Germany Evaluation, 1999). Essentially, microgrids are localized grids that can detach from the conventional grid to function autonomously. Their ability to operate independently of the main grid enhances grid resilience, helping to mitigate grid disturbances and serving as a resource for quicker system response and recovery.

While there is no universally accepted classification in academic or industry circles for categorizing microgrids based on the scale, sophistication of assets, operations, and control functionalities, there are shared characteristics in various definitions. These include (i) the capability to operate in both grid-connected and islanded modes; (ii) clearly defined electrical boundaries; and (iii) the ability to autonomously manage operations by controlling interconnected loads and generation resources.

Microgrids are usually connected to the utility at a campus-wide main metering point and typically interact with the distribution system mainly through bidirectional power flow, especially when selling excess energy back to the grid. Net metering is commonly used to manage this bidirectional power flow, simplifying the compensation process. Ownership structures of microgrids are often straightforward, with utilities considering them as behind-the-meter resources.

There are two primary types of microgrid architectures: i) permanent islands or self-contained, and ii) grid-connected. Self-contained microgrids, often found in remote or completely isolated locations, lack connections to the utility grid. Their generation technologies and infrastructure components (like power electronics, energy storage, and fuel storage tanks) are designed to independently meet the entire load at all times. In contrast, grid-connected microgrids are intended to remain linked to the central grid during normal operations but have the necessary controls and switching capabilities to isolate them from the utility grid during outages. These grid-connected microgrids are typically sized to power only essential loads during utility grid outages, as they usually operate in grid-connected mode.

VPPS

A Virtual Power Plant (VPP) is essentially a network of independent Distributed Energy Resources (DERs) that are coordinated through a cloud-based control system to function as a single large-capacity energy source. It is effectively a distributed power plant created by combining the capacities of diverse DERs. The primary role of a VPP is to coordinate a variety of distributed DERs, enabling their participation in the wholesale markets. These DERs can include dispatchable and non-dispatchable sources, controllable or flexible loads, photovoltaics (PVs), energy storage systems, backup generators, biomass, and small-scale wind turbines, among others.

VPPs are advancing into real-world applications. For instance, in 2019, the South Australian government, in partnership with Tesla, announced a VPP project targeting 50,000 homes (United National Climate Change, 2015). The general goal of VPPs is to facilitate the involvement of small-scale DERs in the wholesale markets, which often have stringent requirements for the availability and reliability of the flexibility offered. VPPs allow the collective trading of power and flexibility from a mix of assets (including power generators and flexible loads) in profitable segments of the wholesale markets, such as balancing reserves. Thus, VPPs are designed with the aim of linking decentralized units under a unified, centralized control system.

Notably, VPPs are often initiated by utilities or governments, with capital investments required for establishing the necessary digital infrastructure. While the data exchange between the VPP control system and individual assets typically uses public communication infrastructure, other digital components like the VPP control software and measurement and control sensors on individual assets still require capital investment. This setup positions these VPPs within the top-down approach to power system management. However, our research seeks to redefine this concept. We aim to design a framework where VPPs not only enable DER participation in wholesale markets but also serve as enablers for localized and decentralized peer-to-peer energy markets. In this context, the operational objective of a VPP extends beyond just maximizing prosumer savings in wholesale markets; it also encompasses interfacing with localized peer-to-peer markets. Thus, VPPs act as a bridge between bottom-up and top-down approaches, with a primary focus on supporting prosumer-centric decision-making. This concept is further elaborated in section 3.

VPP AND MICROGRID: THE DIFFERENTIATION

Virtual Power Plants (VPPs) and microgrids both incorporate a variety of Distributed Energy Resources (DERs), including storage and flexible loads, yet they differ significantly in several aspects:

- VPPs do not possess physical assets; they primarily function within the market and cyber levels, aggregating and efficiently dispatching DERs from various owners. In contrast, microgrids can operate in both grid-connected and islanded modes, maintaining functionality independently in the latter.
- VPPs can integrate assets from any part of the grid, whereas microgrids are typically confined to specific locations, like islands or campuses.
- From the perspective of grid operators, VPPs are considered front-of-the-meter power plants, meaning the resources within VPPs are metered individually. On the other hand, microgrids are often viewed as

behind-the-meter (BTM) resources, with all DER assets of the microgrid metered at a single point of interconnection with the main grid.

In our proposed framework, we further delineate the operational objectives and functionalities of these two aggregation mediums. This distinction is aimed at enhancing the design of prosumer-centric decentralized P2P energy markets that coexist with the utility grid in the future smart grid. These differences are elaborated in section 3.

P2P ENERGY MARKETS

Peer-to-Peer (P2P) markets, also known as Transaction Energy (TE) markets, enable small-scale suppliers to compete with traditional providers of goods or services. Their core purpose is to facilitate the discovery and transaction process between buyers and sellers of varying scales, ensuring trustworthiness and convenience. Similarly, P2P energy markets allow prosumers to directly exchange the energy they generate on-site with their neighbors or peers.

P2P markets serve as a tool for decentralizing energy markets by enabling electricity trading without the need for intermediaries like a central authority or organization. However, P2P energy markets are distinct in their trade of electricity—a commodity that is challenging to store in large quantities and needs to be available on demand. Bulk storage of electricity is neither practical nor feasible, and it is impossible to have customers wait in line for electricity as they would for a physical product. Complicating matters further is the intermittent nature of renewable energy sources and the constantly fluctuating demand.

Due to these unique aspects of electricity as a commodity, designing P2P energy markets is a complex task. It requires careful consideration of both the market and physical levels of the system, as well as the timescale axis. Incorporating sophisticated algorithms for dynamic pricing, real-time balancing of supply and demand, and ensuring grid stability is essential. Advanced metering infrastructure and blockchain technology are also crucial for transparent, secure, and efficient energy transactions. Moreover, grid-responsive demand-side management strategies and intelligent distribution network operations must be integrated into the system. These technologies aid in optimizing resource allocation, reducing peak load, and mitigating the variability and unpredictability of renewable energy sources.

Different spheres operate simultaneously to make the P2P Energy Market work, as shown in Figure 6.1 below.

Additionally, for P2P energy markets to effectively coexist with the existing centralized electricity market infrastructure, their interactions with

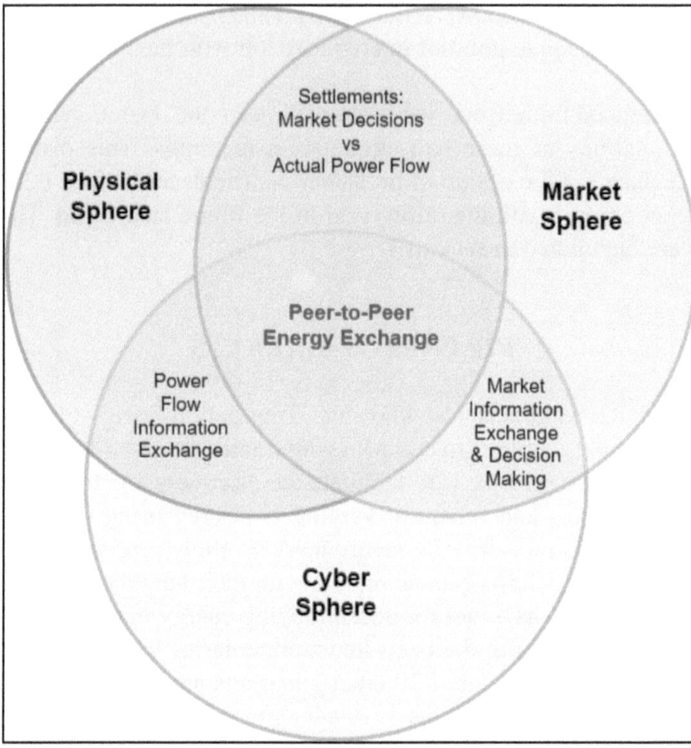

Figure 6.1. Operational Sphere of P2P Energy Markets. Source: Created by the authors of this chapter.

Source: Created by the authors of this chapter.

wholesale markets must also be taken into account in the framework's design. This involves establishing regulatory policies, market rules, and technical standards that support the coexistence of centralized and decentralized market structures. It also necessitates a robust communication and data exchange network, providing a seamless flow of information between distributed energy resources, grid operators, and market participants.

PROPOSED FRAMEWORK: BLUEPRINT AND OPERATIONAL MECHANICS

Existing Obstacles and Literature Deficiencies

The increasing focus on empowering a bottom-up approach for managing and operating the electricity grid necessitates the development of strategies to minimize friction between the long-standing top-down and the emerging bottom-up approaches to grid operations and management. While numerous studies have addressed peer-to-peer (P2P) energy market design and Virtual Power Plant (VPP) operations, a comprehensive systems approach that incorporates existing industry players (such as wholesale markets and load-serving entities) and accounts for their inevitable interactions with the central grid (top-down operations) is still lacking. This involves crafting P2P market operations in a way that aligns the goals of market participants (prosumers) with those of upstream entities (utilities, load-serving entities, aggregators, etc.).

Furthermore, it is crucial that prosumers are not hindered by top-down economic constraints from maximizing their individual cost savings through the operation of their Distributed Energy Resources (DERs). The primary aim of our bottom-up approach is to maximize prosumer benefits without being restricted by these constraints, while also considering the physical limitations of the power systems infrastructure. This approach, which prioritizes prosumer benefit maximization and incorporates ancillary market participation as a secondary objective, is novel in the literature to the best of our knowledge.

Scalability is also a key factor in the effective design of P2P markets. As the number of participants in a market increases, system performance can be adversely affected. A viable solution to this challenge is to segment markets into localized modules or units, which can then be interconnected in a hierarchical structure, allowing different P2P markets to operate at various levels. This structured compartmentalization, which we incorporate into our framework, provides scalability potential.

The challenge in establishing such interconnected hierarchical energy markets lies in ensuring interoperability across the physical, market, and cyber domains. Physical interoperability is partly achieved through interconnected distribution lines, facilitating power flow between different market spheres. Market interoperability involves designing trade mechanisms that allow for efficient transactions within and across different market spheres.

Cyber interoperability, which involves secure information exchange for power flow and market mechanisms, is beyond the scope of this work. To navigate the dynamic landscape of decentralized energy systems, it is essential to integrate advanced optimization algorithms and machine learning techniques into P2P energy market designs. This integration will enable the use of predictive analytics and real-time data processing, deploying intelligent agents within the P2P markets for autonomous decision-making in response to grid changes.

In the context of top-down economic constraints, sophisticated demand response mechanisms and advanced control strategies are necessary for optimal DER utilization by prosumers. The integration of smart grid technologies, such as real-time demand forecasting and grid state estimation, is crucial for ensuring a seamless interaction between the bottom-up and top-down approaches to grid operations. The proposed framework, through intelligent agents and advanced algorithms, ensures that prosumers can fully utilize their DERs within the framework of globally cost-effective grid operations, striking a balance between individual savings and system-wide efficiency.

UNIFIED HIERARCHICAL FRAMEWORK: EMPIRICAL CONSTRUCTION

The central idea of our proposed design is to reimagine the functioning of Distributed Energy Resources (DERs) and the role of prosumers, emphasizing a bottom-up approach to the management and operation of the electricity grid. This approach is not intended to be in stark contrast to the top-down approach. Our framework aims to integrate various smart grid concepts like DERs, microgrids, and Virtual Power Plants (VPPs), aligning their operational goals to foster prosumer-centric Peer-to-Peer (P2P) energy markets. These markets are designed to be compatible with wholesale markets, offering prosumers the flexibility to provide their DER services at the wholesale level.

Transactive Energy (TE) marks a significant paradigm shift in the smart grid sector, transforming end-users from passive consumers to active participants. This concept broadens the scope and inclusivity of user roles within the smart grid, enabling a range of interactive opportunities. TE is centered around enabling dynamic, bidirectional energy and information exchanges, allowing end-users to not just consume but also produce, store, and manage energy more cohesively and efficiently. Utilizing advanced technologies such as the Internet of Things (IoT), Artificial Intelligence (AI), and blockchain, TE promotes a more decentralized, resilient, and sustainable energy system. This approach greatly improves grid reliability, optimizes resource use, and empowers consumers by giving them more control over their energy consumption and production, thereby driving a fundamental shift in the energy sector.

A crucial aspect of truly decentralized energy system operations is prosumer-centric decision-making. This means prosumers have the autonomy to make informed decisions about the utilization of energy produced or stored on-site. In this model, prosumers should have access to a range of options at any given time, as illustrated in Figure 6.2.

When prosumers have the liberty to select from the aforementioned options, they can strategize their operations to minimize energy costs while adhering to their consumption habits and preferences. To establish such a

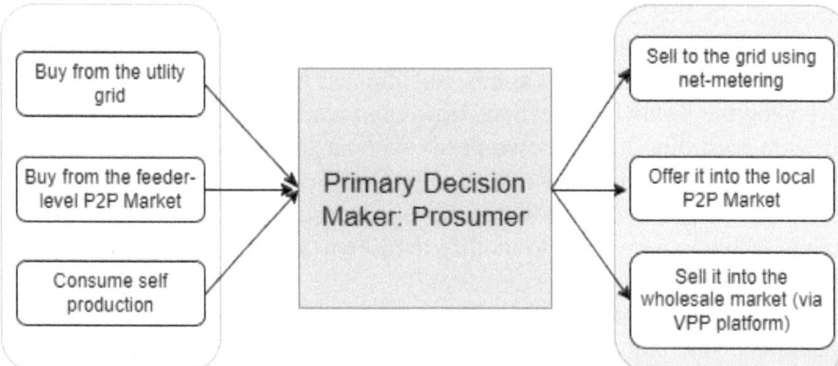

Figure 6.2. Proposed Framework Is Prosumer-Centric at Its Core: At Any Given Time, Prosumers Have Multiple Choices to Buy or Sell Energy.[AQ6]
Source: Created by the authors of this chapter.

prosumer-focused model in the electricity sector's operational planning, it is necessary to reconsider the traditional top-down approach to operational planning, allowing room for bottom-up strategies. This concept is further elaborated in the subsequent paragraphs.

Traditionally, power system planning begins with capacity planning, which involves the development of utility-scale generation, transmission, and distribution infrastructure over a 3–10-year horizon. Capacity Markets primarily function at this level, compensating generation entities that enter into binding agreements with Independent System Operators (ISOs) to provide future generation capacity. Following capacity planning, operational planning looks months ahead to ensure reliable supply while considering major planned outages, such as maintenance and upgrades of centralized power plants, high-voltage transmission lines, or transformers. Bilateral transactions between large-scale market players, conducted without ISO/RTO mediation, occur within this timeframe. In these transactions, the identities of the parties are known to each other. The Financial Transmission Rights Market, which books transmission capacities to facilitate energy flow committed in bilateral transactions, also operates on this horizon.

Energy Markets are then planned and executed on a day-ahead (unit commitment) and hour-ahead (economic dispatch) basis. As real-time approaches, Ancillary Markets come into play to guarantee supply reliability. The fundamental requirement of power systems operations is to maintain a supply-demand balance on a micro-second, near-real-time basis. This balance can be achieved by adjusting either generation or load. The generation-side adjustment mechanisms are discussed in section 4. On the load-side (or demand-side), the traditional top-down approach incentivizes prosumers to alter their consumption patterns for monetary benefits, a process known as Demand

Response. In this scenario, Distributed Energy Resources (DERs), which include flexible loads, are summoned to reduce consumption to maintain the supply-demand balance. It is at this juncture that we propose a paradigm shift by posing the following question: How can prosumers optimize their DERs' dispatch according to their own needs without losing the financial benefits that the top-down demand response program provides through economic incentives? Instead of following the operational decisions imposed from above, which require them to modify their own schedules, why can they not choose their own dispatch preferences?

The question posed is central to the concept of a bottom-up approach in grid operations. Had this question been asked a decade earlier, it might not have been as relevant due to the uncertain future of Distributed Energy Resources (DERs) at that time. However, in today's context, with the increasing deployment of DERs, the decentralization of power systems is increasingly seen as the most viable route towards a low-carbon future. The essence of decentralization lies in empowering individuals with choices and the capability to make decisions based on those choices. Consequently, we suggest that the operational planning of DERs, owned by prosumers, should be approached in a manner akin to traditional top-down operational planning, but facilitated through mechanisms like Peer-to-Peer (P2P) markets and non-profit digital aggregators such as Virtual Power Plants (VPPs).

In our approach, we don't suggest replacing the top-down approach with a bottom-up one. Instead, we propose a *coordinated operational planning scheme that allows for the coexistence and coordination of both approaches*. This is illustrated in Figure 6.3 (Mishra et al., 2023). The first inverted triangle in the figure represents the current power systems planning process, which includes both infrastructure and operational planning across various time horizons. In the existing operations of distribution systems, the second operational planning triangle, which represents the bottom-up approach, is absent. This means that, traditionally, Distributed Energy Resources (DERs) only react to demand response signals from utilities or aggregators. However, in our proposed model, prosumers (like individual microgrids) can plan their operations, including energy selling or purchasing, in advance (up to 30 minutes ahead). They can then decide to participate in demand response programs (of the non-dispatchable type) if it proves cost-effective, using the Virtual Power Plant's (VPP's) aggregation platform (15 minutes ahead).

The subsequent sub-section provides a detailed description of this operational scheme across all three levels of the hierarchical framework.

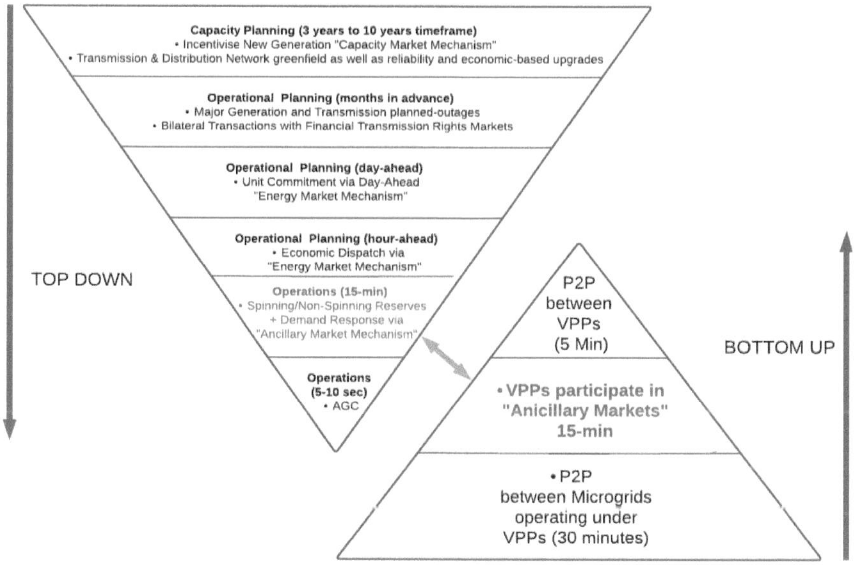

Figure 6.3. Highlight of Our Proposed Framework: Marriage of Top-Down and Bottom-Up Approaches to Electricity Grid Operations.

Source: Created by the authors of this chapter.

UNIFIED HIERARCHICAL FRAMEWORK: COORDINATED OPERATIONS PLANNING

In our outlined framework, the lower triangle from Figure 6.3 represents the hierarchy of dispatch decisions, beginning with the individual prosumers' choices, as shown in the prosumer choice model in Figure 6.2. In Figure 6.3 illustration, which combines top-down and bottom-up approaches to electricity grid operations, this represents the lowest level of the bottom triangle. At this level, prosumers (represented as microgrids in our model) engage in the 30-minute ahead Peer-to-Peer (P2P) energy markets at the feeder level. Virtual Power Plants (VPPs) serve as a semi-autonomous digital platform, facilitating interactions among prosumers. These VPPs calculate the surplus from the 30-minute ahead P2P markets and offer this aggregated excess energy into the Ancillary Markets, which operate on a 15-minute ahead timeline (indicated by the green arrow in the diagram). The third level of our bottom-up operational planning scheme involves P2P energy markets between various VPPs, functioning on a 5-minute time horizon. This market essentially acts as a secondary, bottom-up market for the VPPs. If VPPs are unable to deliver the energy they committed to in the 15-minute ahead Ancillary Markets due to fluctuations in the first level (P2P between individual microgrids), they engage in bilateral transactions among themselves to fulfill their commitments in the Ancillary Markets. This operational planning scheme[4] is depicted in Figure 6.4 (Mishra et al., 2023).

The operational framework illustrated in Figure 6.4 establishes a resilient and adaptive distribution grid by integrating energy trading and management across three distinct levels, each complementing the others in terms of operational timeframes. This multi-level system facilitates a dynamic equilibrium between supply and demand over various time scales, thereby improving the efficiency of localized, small-scale decentralized markets. Beginning with individual microgrids and extending to interactions among Virtual Power Plants (VPPs), this bottom-up approach offers a decentralized yet coordinated method for conducting peer-to-peer trades and marketing excess energy in ancillary markets. This aspect is particularly vital for handling the intermittent nature of renewable energy sources at the microgrid level. The framework's design allows for swift adjustments in response to unexpected demand or supply fluctuations, ensuring reliability and minimizing penalties. Additionally, the inclusion of ancillary markets as a key element in this hierarchical structure underscores the framework's capacity to optimize energy distribution and maintain grid stability, showcasing its effectiveness in navigating the complexities of contemporary energy systems.

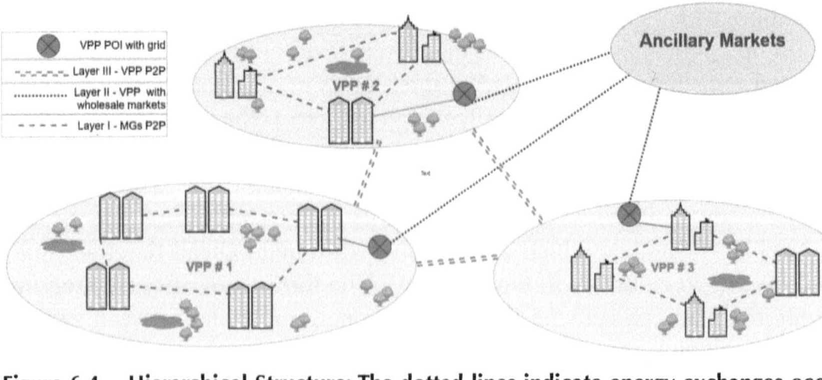

Figure 6.4. Hierarchical Structure: The dotted lines indicate energy exchanges occurring within and between various Virtual Power Plants (VPPs). Additionally, these VPPs are linked to Level II wholesale markets.

Source: Created by the authors of this chapter.

LEVEL I—INDIVIDUAL MICROGRIDS AND THEIR LOCALIZED PEER-TO-PEER MARKETS

In our proposed framework, individual microgrids function as behind-the-meter resources, meaning their energy consumption and injection are measured at the interconnection point between the distribution grid and the microgrid's main meter. The operation and dispatch of the Distributed Energy Resources (DERs) within a microgrid are managed at the prosumer or owner level. The goal of the intelligent microgrid controller is to minimize energy costs by considering various external price signals, including utility energy prices, peer-to-peer (P2P) market buy/sell prices, and compensation for committing energy to Virtual Power Plants (VPPs) for ancillary market purposes, alongside internal consumption patterns and preferences.

These microgrids, referred to as prosumers, participate in Level I P2P energy markets (the lowest level in the inverted triangle of Figure 6.3). They plan their dispatch 30 minutes ahead, making buy or sell offers in the P2P markets, which operate digitally in the cyber sphere. Each feeder, hosting multiple prosumers, conducts its P2P market using bilateral contracts. It is important to note that in the United States, feeder lengths in power distribution networks can vary significantly, with the length often correlating with local load density.

The process of organizing these P2P energy trades is detailed in Figure 6.5 (Mishra et al., 2023), illustrating a complex sequence of events spanning the physical, cyber, and market spheres of energy system operations. This begins with prosumers engaging in the market sphere to place bids, followed by the recording of this data in the cyber sphere. A key aspect of this process, as depicted in Figure 6.5, is that the P2P market sphere at Level I initiates the overall P2P trading mechanism. After the energy transaction at a specific time, the data is recorded in the cyber network, with settlements occurring post facto.

The third step involves power flow analysis, conducted by the autonomous agent of the VPP, separate from the individual prosumer's microgrid control. This analysis is crucial for ensuring feeder and upstream network reliability. The microgrid controller maintains voltage and frequency stability at the Point of Interconnection (POI) through its hierarchical control. However, power flow analysis serves as an auxiliary process, primarily aimed at identifying any potential reliability issues, such as persistent voltage violations, within the feeder. The results from the power flow analysis can trigger alerts, enabling the VPP platform to instruct prosumers to adjust their real and reactive power contributions to the distribution grid's feeder. Notably, the prices at which prosumers buy or sell energy are not determined by the power flow

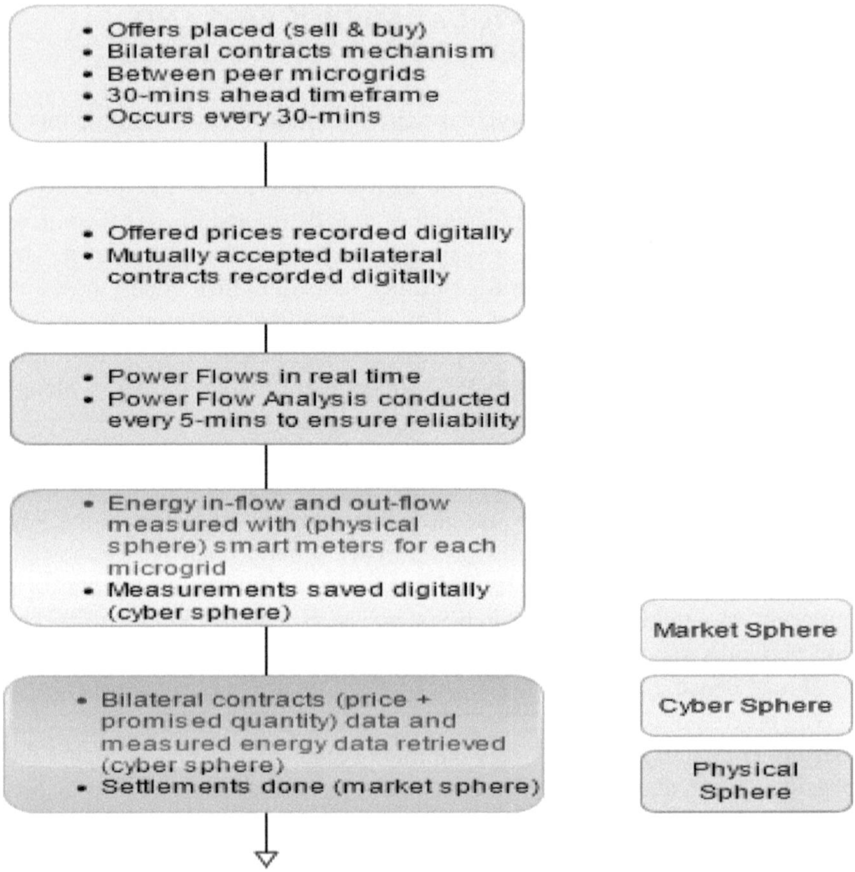

Figure 6.5. Level I—Energy Markets Design and Operations.
Source: Created by the authors of this chapter.

analysis. This arrangement between the market and physical spheres is feasible due to the small-scale nature of energy flows and transactions at the feeder level, which constitutes the Level I P2P energy markets in our framework.

A key aspect of the Level I Peer-to-Peer (P2P) Market design is its largely autonomous operation from the broader upstream market activities. Essentially, this represents a grassroots approach where frontline prosumers have the autonomy to make decisions about their energy production and consumption without direct influence from the larger-scale wholesale or retail market operations. It is important to clarify that this independence pertains to direct influence; indirectly, the pricing in the Level I P2P market is still influenced by upstream market prices, as prosumers retain the option to purchase energy either from the utility or the P2P market. This competitive setup inevitably leads to secondary and tertiary causal effects. However, exploring these causal relationships falls outside the scope of this work.

Consequently, the pricing mechanism within Level I P2P energy markets, which can be considered free markets, can be effectively established through a combination of model predictive control and game theory approaches. In this context, a free market is defined as an economic system where prices are set by open competition among private entities (prosumers in our case).

LEVEL II—VIRTUAL POWER PLANTS MEET ANCILLARY MARKETS

The second tier of our proposed model acts as a crucial link between the overarching top-down and grassroots bottom-up market operations. This level provides a pathway for individual prosumers to engage in wholesale markets, positioning them as "price-setters"—a fundamental aspect of the decentralization concept. The key decision-makers at this level are Virtual Power Plants (VPPs). Defined as semi-autonomous, non-profit digital platforms, VPPs aim to optimize energy cost savings for participants at Level I (referencing the VPPs' ownership structure). It is important to note that the capital investment required for establishing the digital infrastructure of VPPs is borne by the Level I community. This community ownership ensures that VPPs function not as external aggregators seeking profit maximization but as intelligent, semi-autonomous agents representing the Level I community in wholesale markets by aggregating assets owned by community prosumers.

Functionally, Level II primarily manages the surplus or deficit of energy from Level I P2P energy markets, utilizing any excess energy to place bids in the 15-minute ancillary services markets. The operations of the VPP platform in both the market and cyber domains are closely linked, with the VPP continuously monitoring Level I P2P market status to inform its bidding strategies in the ancillary markets. Additionally, the VPP engages in the physical operational sphere by running power flow programs, communicating any identified reliability issues to Level I P2P market participants, and suggesting corrective actions. The operations of the VPP platform are detailed in the flow chart presented in Figure 6.6 (Mishra et al., 2023).

The chosen time horizons for Level I (30 minutes) and Level II (15 minutes) are significant. Level I operates with 30-minute ahead bids for 30-minute intervals, while Level II functions on a 15-minute ahead basis for 15-minute intervals, allowing it to adjust its ancillary market bidding strategies based on the actual energy surplus or shortfall from Level I. This arrangement aids Level I participants in maximizing the value of their energy by selling it in

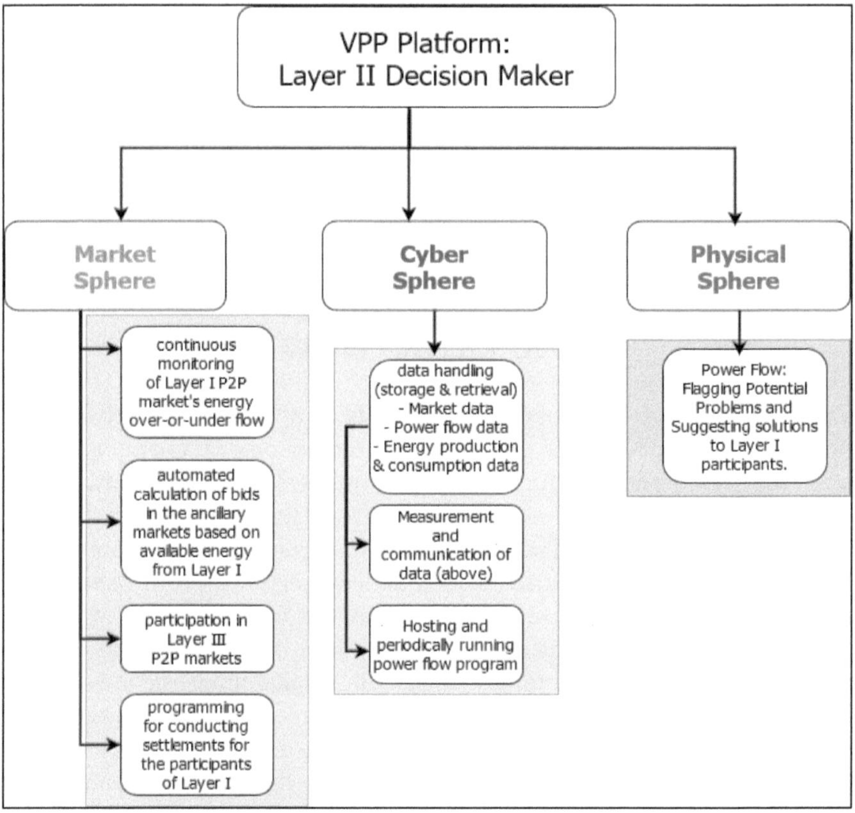

Figure 6.6. Functionalities of the Semi-Automated VPP Platform—Level II Decision Maker.

Source: Created by the authors of this chapter.

the wholesale (ancillary) markets, necessitating a shorter operational time horizon for Level II compared to Level I.

LEVEL III—FEEDER LEVEL PEER-TO-PEER ENERGY MARKETS

Level III facilitates P2P energy markets among VPPs, operating on the most immediate time frame of 5 minutes. The primary role of these Level III P2P markets is to balance the surplus or shortfall emanating from Level II. Specifically, if a VPP has committed to supplying a certain quantity of energy in the ancillary market for a future 15-minute slot and subsequently realizes that a shortfall from the Level I P2P energy market will hinder its ability to fulfill this commitment, it can then acquire the required energy by engaging in the Level III P2P market. Essentially, Level III P2P markets function as a secondary market for VPPs, enabling them to purchase energy in 5-minute increments to circumvent potential penalty charges imposed by the ancillary markets. The operational dynamics of all three levels in the framework are systematically outlined in Figure 6.7.

MERITS OF THE PROPOSED FRAMEWORK

In an era where sustainable energy solutions and efficient grid management are paramount, the proposed framework emerges as an innovative model in the field of smart grid technology. It represents a paradigm shift from conventional top-down energy management approaches to a more dynamic, consumer-centric model. This is achieved by leveraging peer-to-peer (P2P) energy markets, distributed energy resources (DERs), and virtual power plants (VPPs) within a hierarchical operational structure. The framework's design is unique in its emphasis on prosumer engagement, decentralization of energy system operations, and its novel approach to handling the complexities of modern energy grids. This holistic approach not only empowers consumers but also ensures compatibility with existing grid infrastructure and regulatory frameworks.

Tasks	Time Scale	Operational Objectives	Functionalities	Ownership Structure
Level I (Microgrids P2P Market)	30-mins ahead (in 30-min increments)	• Cost-minimization • Maintain Prosumer Preference	• MG internal energy management • Intelligent cyber layer designed to find optimal buy-sell decisions	All the assets (DERs, flexible loads, energy storage) are owned by the behind-the-meter prosumer
Level II (VPP /aggregators participate in Ancillary Markets)	15-mins ahead (in 15-min increments)	• Manage the "overflow" of net positive or net negative energy from the Layer I • Offer the extra energy into the ancillary markets	• Simultaneously work with Layer I and Layer III with the objective of maximizing profits of Layer I participants	VPP as a "digital entity" owned by the community. May or may not own a "community battery"
Level III (VPP P2P Markets)	5 mins ahead (in 5-min increments)	• Manage the over-or-under flow from Layer II through bilateral contract mechanism	• Serve as "secondary markets" for VPPs to buy or sell energy to meet the promised quantity in the ancillary markets	No asset ownership

Figure 6.7. Hierarchical P2P Energy Markets—Operational Details of the Three Levels.
Source: Created by the authors of this chapter.

The framework is structured with a focus on the timescales of the bottom-up operational scheme, ensuring that decision-making cascades from the lowest level upwards. Level I operations are planned 30 minutes ahead, prioritizing prosumer autonomy in operational decisions, thereby reducing reliance on top-down usage modification signals (like demand response notifications sent 5 minutes in advance). The VPP, as a non-profit, semi-autonomous intermediary, functions on a 15-minute timescale to manage the energy surplus or deficit from Level I (where P2P energy markets operate on a 30-minute timescale). Level III operates on a 5-minute timescale, aiding VPPs in fulfilling their commitments in the ancillary markets by addressing the energy imbalances from Level II.

While the framework is primarily tailored for prosumer preferences and cost efficiency, it aligns well with traditional top-down distribution grid operations. It integrates key smart grid components (microgrids, DERs, VPPs, etc.) and redefines their objectives and functionalities to support a bottom-up smart grid approach. This design also facilitates the seamless participation of DERs in wholesale markets, aligning with FERC Order 2222. Key characteristics of the framework include:

- Offers a viable strategy for decentralizing energy system operations, addressing energy equality.
- Establishes a comprehensive decentralized operational model, harmonizing diverse smart grid technologies like DERs, microgrids, and VPPs for a prosumer-focused bottom-up approach.
- Clarifies the distinct roles and interactions of market, physical, and cyber domains within P2P market operations.
- Effectively coordinates centralized and decentralized energy markets, using VPPs as a bridge.
- Models VPPs as semi-autonomous, non-profit entities, enhancing prosumer benefits.
- Scalability: The hierarchical design enables efficient scaling of the proposed P2P energy markets.
- Interoperability: Ensures smooth operational planning across different levels in the hierarchy, avoiding potential timescale conflicts.
- Cost-Effectiveness: Offers economic benefits to both prosumers and the overall grid. By optimizing energy distribution and reducing transmission losses, the framework allows for a more efficient use of resources, leading to lower operational costs and potential savings for consumers.

MAKING THE ECONOMICS WORK

The integration of Peer-to-Peer (P2P) energy markets into the broader energy ecosystem presents a compelling economic proposition for prosumers, who are at the heart of this evolving landscape. By investing in Distributed Energy Resources (DERs), prosumers can effectively participate in these markets, not just as consumers of energy but as active players contributing to a more dynamic, equitable, and decentralized energy system.

One of the primary incentives for prosumers to invest in DERs stems from the economic benefits of participating in P2P energy markets. Unlike traditional models where prosumers sell their surplus energy back to the grid at predetermined rates, often lower than the retail electricity prices, local P2P energy markets—*connected with ancillary wholesale energy markets*—offer a more lucrative avenue for the prosumers. In these markets, prosumers can trade electricity directly with peers, setting prices that are mutually beneficial and often more favorable than the conventional buy-back rates offered by the grid. Moreover, the hierarchy set in Level I, Level II, and Level III has been working in favor of prosumers by increasing the choices/options prosumers have in terms of buying or selling energy.

This direct trading mechanism not only provides additional revenue streams for prosumers but also enhances their autonomy over the energy they produce. By eliminating the need for intermediaries in the transaction process, prosumers avoid the costs associated with third-party coordination of their energy trades. This aspect is particularly significant in the proposed framework, where Level II is modeled as a semi-autonomous, all-digital entity. This design ensures that the prosumers do not incur additional expenses in paying a third party to facilitate their P2P energy transactions. The economic benefit here is twofold: maximizing returns from their investment in DERs and minimizing transaction costs.

Moreover, the proposed framework supports a sustainable economic model for decentralized energy systems. By allowing prosumers to engage directly in energy trading, the framework promotes a more equitable distribution of economic benefits across the energy value chain. This direct engagement also encourages further investment in renewable energy sources, as prosumers see tangible economic returns from their environmentally friendly initiatives.

Therefore, the proposed P2P energy market framework offers a compelling economic case for prosumers. By providing an efficient platform for direct energy trading, encouraging the adoption of renewable energy sources, and eliminating costly intermediaries, the framework ensures that the economics work in favor of those at the forefront of this energy revolution. This approach not only empowers prosumers but also lays the groundwork for a more resilient, efficient, and sustainable energy system.

REGULATORY INCENTIVES

Regulatory incentives are vital in accelerating the transition to decentralized energy systems with a high penetration of clean energy sources. They play a crucial role in lowering the financial risks associated with investing in renewable energy technologies. By making these technologies more accessible, they enable a broader range of investors and consumers to participate in the clean energy market. These incentives also spur innovation and development in the renewable energy sector. They address the critical market barriers such as the high initial costs of renewable installations and the tendency to undervalue the environmental benefits of clean energy sources. This facilitation is crucial in shifting the market dynamics towards more sustainable and eco-friendly energy solutions.

Moreover, regulatory incentives are instrumental in promoting energy independence and security. They help reduce reliance on imported fuels, thereby enhancing national energy security. This shift is not only strategic from an energy standpoint but also contributes to economic stability. The economic impact of these incentives is also significant. They drive job creation in the renewable energy sector and stimulate economic growth by fostering new industries and technologies. This economic stimulation is essential for the long-term sustainability and viability of clean energy solutions.

In terms of environmental impact, these incentives directly contribute to the mitigation of climate change. They support the reduction of greenhouse gas emissions, aligning with global efforts to combat environmental degradation. Finally, the role of regulatory incentives in creating public awareness and acceptance cannot be understated. Financial incentives often raise public consciousness about the benefits of renewable energy, thereby increasing acceptance and adoption rates. This public buy-in is crucial for the widespread implementation of decentralized, clean energy systems.

In both the United States and the European Union, there are several regulatory incentives aimed at encouraging prosumers to install Distributed Energy Resources (DERs) and participate in renewable energy production and peer-to-peer (P2P) energy markets.

In the United States, there are various federal government tax credits, grants, and loan programs available for qualifying renewable energy technologies and projects. These include the Renewable Electricity Production Tax Credit (PTC), the Investment Tax Credit (ITC), the Residential Energy Credit, and the Modified Accelerated Cost-Recovery System (MACRS).

Additionally, most states offer financial incentives to encourage renewable energy production and use. Renewable portfolio standards (RPS) are also in place in many states, requiring a percentage of electric power sales to come from renewable energy sources. Renewable energy certificates or credits (RECs) are financial products that can be traded and are available for sale or purchase. These allow a purchaser to pay for renewable energy production without directly producing or purchasing the energy. Net metering policies, which are available in 44 states and the District of Columbia, allow electric utility customers to install qualifying renewable energy systems and connect them to the grid, billing the customers for the net electricity they use. Feed-in tariffs (FITs) are established in several states and utilities, offering special rates for purchasing electricity from certain types of renewable energy systems, typically higher than standard electricity rates.

In the European Union, the incentives vary by country but are generally designed to promote the adoption of renewable energy technologies and to support prosumers in their role in the energy market. For instance, in Germany, there are feed-in tariffs for renewable technologies, financial subsidies, and tax exemptions for electric vehicle (EV) owners, as well as subsidies for the installation of home battery storage systems. This support allows prosumers to avoid high network fees and reduce their electricity costs. Irish legislation promotes the adoption of EVs and small-scale renewable systems like wind and photovoltaic (PV) systems, offering tax reductions, feed-in tariffs, and financial incentives for households and businesses. Lithuania has set an ambitious goal of turning 50% of its energy consumers into prosumers by 2050, facilitated by changes in net-metering legislation and the introduction of flexible options for acquiring solar power plants and renewable energy.

Regulatory incentives play a pivotal role in creating an environment conducive to the growth of decentralized energy systems. These incentives not only ensure economic viability but also the sustainability of clean energy sources, rendering them essential in the global shift towards a more sustainable energy future. As leading nations worldwide spearhead the establishment of incentivization mechanisms through regulatory frameworks, they send a resounding message of unity. This collective approach significantly reflects their commitment to addressing the pressing issue of global climate change, demonstrating a unified front in the quest for environmental preservation and energy transformation.

As the landscape of energy production and consumption undergoes a significant shift, the focus turns towards innovative technologies that can support and enhance these evolving paradigms. One such technology that stands at the forefront of this transformation is blockchain. Bridging the gap between regulatory incentives and the practical application of decentralized energy systems, blockchain emerges as a critical tool in the realm of P2P energy markets. This transition from policy-driven incentives to technological implementation represents a pivotal step toward realizing the vision of a decentralized, efficient, and sustainable energy future. In the next section, we delve into the role of blockchain in reshaping the dynamics of energy markets, highlighting how this technology not only complements but also amplifies the effectiveness of regulatory incentives in making the economics work as well as fostering a robust and resilient energy ecosystem.

BLOCKCHAIN FOR P2P ENERGY MARKETS

Blockchain technology, a revolutionary digital ledger system, is increasingly becoming a cornerstone in the evolution of P2P energy markets. At its core, blockchain offers a decentralized, transparent, and secure platform for transactions, making it an ideal framework for the energy sector's transformation. In P2P energy markets, where individuals can buy, sell, or exchange energy without traditional intermediaries, blockchain technology enables efficient, trustless transactions. This innovative application not only democratizes energy distribution but also paves the way for a more sustainable and consumer-centric energy landscape, aligning seamlessly with the rising trend of distributed energy resources and renewable energy systems.

Blockchain technology has emerged as a transformative force in the domain of P2P energy trading networks, introducing several advantages that promise to redefine the energy market landscape. A key benefit is the decentralization and efficiency it brings to these networks. Blockchain technology facilitates direct trading between energy consumers and producers, eliminating the need for intermediaries. This not only streamlines the process but also enhances overall market efficiency. Another significant advantage lies in its ability to bolster security and privacy. Blockchain provides a secure environment for energy trading, adeptly addressing the challenges prevalent in scenarios like microgrids and energy harvesting networks. Furthermore, the technology introduces a high level of automation and enables fast, real-time settlements through smart contracts. This aspect of blockchain not only improves the fairness of energy trading but also fortifies its security, making it a robust solution for modern energy challenges.

Despite these advantages, blockchain technology in P2P energy trading is not without its drawbacks. A notable concern is the environmental impact associated with blockchain mining, as it entails significant energy consumption and carbon emissions. This aspect raises substantial environmental concerns, particularly in the context of global efforts to reduce carbon footprints. Another disadvantage is the cost implications associated with the technology. The implementation of blockchain in energy trading necessitates substantial computational resources, which in turn incurs considerable costs. Lastly, the complexity of designing and implementing a decentralized P2P market is a formidable challenge. Balancing economic efficiency with information privacy requires intricate and sophisticated design.

While blockchain technology offers numerous benefits such as enhanced decentralization, security, and automation in P2P energy trading networks, it is imperative to consider the challenges it presents. These include its environmental impact, the costs involved, and the complexity of its implementation. As this technology continues to evolve, addressing these challenges will be crucial for realizing its full potential in transforming the energy trading sector.

CONCLUSION

Originally configured for a one-way power flow from centralized power plants, the grid is now adapting to a two-way flow due to the integration of DERs, which feed power back into the system. Consumers, traditionally passive in their electricity usage, are evolving into "prosumers"—actively managing their energy consumption, production, and storage. This shift is altering the traditional dynamics of the energy market, ushering in a new era of consumer empowerment and involvement in energy-related decision-making. The growth of smart cities and the trend of urbanization are additional catalysts in this transformation. Smart cities, with their integration of technology into various infrastructures, including energy, are pivotal in enhancing efficiency, sustainability, and livability.

DERs, often categorized as behind-the-meter (BTM) sources, are predominantly installed by commercial, industrial, and residential customers, independent of utilities or regulatory bodies. This localized generation, synonymous with DERs, brings several benefits to the evolving energy system, including i) minimizing losses from long-distance power transmission; ii) incorporating more renewable energy sources like rooftop solar, thereby reducing carbon emissions; iii) enhancing resilience by maintaining critical loads during grid outages; iv) reducing the need for upstream generation and transmission capacity, which is a significant investment in both aging and developing grids; v) decentralizing electricity production and supply, reducing societal dependence on utility grids, which are traditionally the sole power source for most people.

While localized generation offers numerous advantages, encouraging end-users to invest in DERs necessitates that their benefits surpass the costs. Additionally, traditional utility revenue models, based on a unidirectional energy flow, are not fully compatible with the concept of DERs selling power back to the grid. To adapt to the growing trend of DER deployment, utilities are rethinking their business models, which were previously dependent on the conventional energy flow chain. From a technical standpoint, the unpredictable and dispersed nature of DERs presents challenges for reliable and efficient grid operation. Historically, grid mechanisms were designed for centralized, large-scale power plants, distributing electricity via high-voltage transmission networks and ensuring supply reliability with fossil-fuel-based generation to meet fluctuating demand. However, the variability and unpredictability of renewable energy-based DERs make the traditional operation mechanisms less effective.

Despite the challenges associated with integrating localized renewable energy generation into the central grid, the significant benefits it offers make it essential to address these issues. Therefore, the path forward involves creating and applying both business and technical operational models that leverage the strengths of DERs while mitigating their limitations. This is the essence of this work—the development of a framework for bottom-up energy systems operations that effectively integrates with the existing top-down operational approach. This framework is designed to capitalize on the synergies between these two grid management strategies by coordinating the dispatch decision timelines of various network agents. Its primary goal is to place prosumer decision-making at the forefront, while also ensuring benefits for centralized grid operators, particularly when VPPs engage in the ancillary services markets.

The framework outlined in the chapter is designed to create a prosumer-focused model for grid operations, effectively merging top-down and bottom-up approaches in electricity market operations and management, with a strong emphasis on prosumer savings and preferences. The practical implications of this integration include:

- Enhanced Prosumer Autonomy: Prosumers gain increased control and flexibility over their distributed energy resources (DERs). They can choose how to utilize the energy they generate or store, whether it is for consumption, storage, or sale.
- Participation in Multiple Markets: Prosumers can engage in both local peer-to-peer (P2P) energy markets and larger centralized wholesale markets. This dual participation opens up various avenues for revenue generation and value creation. Importantly, the framework allows prosumers to autonomously or algorithmically select the most profitable market options, acting as unbiased digital agents in these transactions.
- Decentralization and Democratization of the Energy Sector: The proposed model fosters a more decentralized and democratic energy landscape. Prosumers are not confined to being mere price-takers in utility-driven schemes like net metering or dependent on third-party

Virtual Power Plant (VPP) providers for selling their energy. Instead, they are empowered to make independent decisions, reducing their reliance on large-scale utilities and intermediaries. This shift promotes a more balanced and equitable energy market where the role and influence of individual prosumers are significantly enhanced.

APPENDIX I

The global electricity sector is undergoing a significant transformation driven by various factors:

1. Climate Change Impacts: Extreme weather conditions, such as intense heatwaves and cold spells, are becoming increasingly common due to climate change. This has escalated the urgency to reduce the carbon footprint of energy systems. The electricity sector, being a major contributor to global carbon emissions, is under pressure to shift towards cleaner, renewable energy sources. This transition is crucial not only for mitigating climate change but also for adapting to its impacts, ensuring a more sustainable and environmentally friendly energy future.
2. Growing Global Energy Demand: The planet's energy needs are expanding rapidly, fueled by population growth, urbanization, and the economic development of emerging nations. This surge in demand presents a significant challenge: how to sustainably meet this increasing energy requirement. The solution lies in developing and deploying energy systems that are efficient, scalable, and capable of supporting the growing energy needs without exacerbating environmental issues.
3. Aging Infrastructure in Developed Countries: In many developed countries, the central electricity grids are aging and require substantial capital investment for upgrades. These upgrades are essential to maintain the reliability and efficiency of the power supply. As these grids age, they become more susceptible to failures and less capable of meeting modern energy demands, necessitating significant investments in modernization and refurbishment.
4. Need for Resilience Against Natural Disasters: The rise in low-probability, high-impact events like hurricanes and earthquakes has highlighted the need for more resilient power grids. These grids must be capable of sustaining critical loads during such catastrophic events to ensure uninterrupted power supply, especially in crucial areas like hospitals and emergency services.
5. Technological Advancements: Several technological developments are reshaping the electricity sector:

- Large-scale DER Deployment: The increased production scales and falling costs of Distributed Energy Resources (DERs) are leading to their wider adoption. This shift is decentralizing energy production and enabling more sustainable and localized energy solutions.
- Consumer-Level ICT and IoT: The proliferation of consumer-level Information and Communications Technologies (ICT) is ushering in an era of hyper-connectivity, marked by the Internet of Things (IoT). This connectivity is transforming how energy systems are monitored, controlled, and optimized.
- Measurement and Sensing Infrastructure: The decreasing costs of sensors have led to the widespread deployment of measurement and sensing infrastructure. This development is generating vast amounts of data, offering unprecedented insights into grid operations and energy usage patterns.
- Electric Vehicles (EVs): The growing number of EVs on the roads, with even higher adoption anticipated in the future, is changing the dynamics of energy demand and distribution.
- Grid-Responsive High-Performance Buildings: The increase in buildings equipped with advanced analytics and controls is leading to the development of grid-responsive, high-performance structures that can efficiently manage energy consumption and contribute to grid stability.

APPENDIX II

Who's Who in the Electricity Sector—*Agents* in Energy Systems:

In the electricity sector, understanding the roles and organization of various agents is crucial for designing effective market operations. These agents play key roles in the journey of energy from generation to end-users:

- Electric utilities: These are either public or private entities responsible for owning and operating the infrastructure for transmitting and distributing electricity. They ensure the supply of electricity to both households and businesses.
- Vertically integrated utilities: These utilities not only own the transmission and distribution networks but also the power generation plants. They manage every aspect of electricity production and supply in their service areas.
- Load-serving entities (or local distribution companies): These entities are responsible for delivering electric power to end-users. Depending on the region, LSEs may generate their own electricity or simply manage

the grid infrastructure needed to deliver electricity from other sources to customers.

- Electricity retailers: In areas with deregulated electricity markets, these retailers sell electricity directly to consumers. Here, the local distribution companies own the distribution networks and handle the physical delivery of electricity.
- Grid operators or transmission system operators (TSOs): These operators are essential for balancing the grid. They ensure that the electricity supplied into the grid matches the consumption by users, maintaining grid stability. They coordinate with utilities, generators, and retailers to prevent issues like blackouts or equipment damage due to imbalances in power supply.
- Electric Utilities as Grid Operators: In some regions, electric utilities also function as grid operators. However, in other areas, this role is fulfilled by Regional Transmission Organizations (RTOs) or Independent System Operators (ISOs). Post-deregulation, RTOs have often taken over the role of grid operators from utilities and now manage wholesale electricity markets. Unlike utilities, RTOs and ISOs operate the grid but do not own the underlying resources and infrastructure, such as power plants and transmission lines.

APPENDIX III

Wholesale Markets and Demand Response

In the United States, three types of market mechanisms operate in conjunction:

Energy Markets

Energy markets function as auctions to coordinate daily electricity production. In these markets, electric suppliers propose selling electricity generated by their power plants at a specific bid price. Load-serving entities (LSEs), representing the demand side, bid for this electricity to fulfill their customers' energy needs. Offers from the supply side are arranged in ascending order based on the offer price. The market reaches equilibrium when the electricity offered aligns with the demand, and generators are paid the market price for each megawatt-hour produced. Before finalizing the most cost-effective resource set to meet the anticipated daily energy demand, Independent System Operators (ISOs) consider various factors like weather forecasts, generator availability, and transmission line outages. Energy markets are categorized into day-ahead and real-time markets.

The day-ahead market is responsible for scheduling electricity production, ancillary service commitments, and consumption for the following day. It establishes financially binding schedules for electricity generation and consumption one day prior to actual production and use. In these markets, energy supply and consumption schedules are determined hours before the start of the operating day. The RTO/ISO runs a computerized market model that pairs buyers and sellers across the market area for each hour of the day. This model assesses participant bids and offers based on the power flows required to transport electricity from generators to consumers across the grid. It also accounts for variations in system capabilities due to factors like weather and equipment failures, as well as rules and procedures for maintaining system reliability. Bids for generation and demand scheduled in the day-ahead market are settled at the day-ahead market prices.

The real-time market addresses discrepancies between the day-ahead market schedule and actual real-time load while adhering to reliability standards, including unplanned outages and transmission line capacity limits. Operated in five-minute intervals, the real-time market clears a smaller volume of energy and ancillary services compared to the day-ahead market. It also offers additional opportunities for suppliers to offer energy. Differences between real-time generation and load compared to the day-ahead cleared amounts are settled at real-time prices, which tend to be more volatile than day-ahead market prices.

Capacity Markets

Electricity retailers are mandated by the North American Electric Reliability Corporation (NERC), an autonomous entity overseeing grid reliability, to maintain sufficient generating capacity to cover both the anticipated load and a reserve margin for grid stability. Some Regional Transmission Organizations (RTOs) conduct capacity auctions, offering retailers a method to fulfill their capacity needs while allowing generators to recoup fixed costs—expenses that remain constant regardless of electricity production levels—which might not be fully compensated through energy markets alone. Capacity markets aim to financially motivate suppliers to maintain operational generation assets and encourage new generation investments. These markets typically operate as forward markets, ensuring that generation capacity is available and ready to produce electricity at least a year in advance.

The capacity market incentivizes the upkeep of existing generation facilities and the creation of new capacity sources, safeguarding the power system's future reliability. In the capacity market auction, generators bid a price equivalent to the cost of keeping their plant operational and ready if needed. These bids are organized in ascending order of price. When the bids match the collective requirement of all retailers to meet expected peak demand plus a reserve margin, the market reaches equilibrium. At this juncture, all generators that cleared the market receive the same clearing price, set by the bid of the last generator needed to fulfill demand. Capacity markets serve a dual purpose: i) They ensure long-term reliability by securing adequate power to meet consumer needs a few years ahead (typically three years) and ii) They provide payments to generators as a reward for their readiness to operate and supply electricity when required, thus ensuring medium-term supply reliability. However, if generators fail to operate when called upon, they may incur penalties under capacity performance requirements.

Ancillary Markets

Ancillary Services Markets enable ISOs/RTOs to keep a reserve of backup generation for scenarios like unexpectedly high demand or system contingencies, such as generator failures or transformer breakdowns. These markets bolster both the reliability of the electricity supply and its transmission. The services offered in these markets are utilized in real-time or in the immediate future.

There are various types of ancillary services, each corresponding to the urgency with which backup generation must be deployed. The North American Electric Reliability Corporation (NERC) and regional entities set the minimum requirements for each ancillary service to ensure grid stability. These services are broadly categorized into two groups: i) Regulation and ii) Operating Reserves. Regulation services align generation with short-term fluctuations in load by adjusting the output of specific resources, typically every few seconds, through an automatic control signal. This adjustment helps maintain the system frequency at 60 Hertz. Operating reserves, in contrast, are *additional ancillary services activated to rebalance load and generation when a supply resource unexpectedly goes offline*. Operating reserves are further divided into i) Spinning; ii) Non-spinning; and iii) Supplemental reserves, differentiated by the speed at which they can commence electricity supply following dispatch signals.

Wholesale Market Pricing

Due to the deregulated nature of electricity markets, many wholesale markets currently depend on competitive market dynamics to determine prices. It is important to note that not every region or country has wholesale markets managed by Independent System Operators. For instance, in Canada, certain provinces like Manitoba still operate under a traditional vertically integrated model, where a single large provider of bundled electricity services sets prices based on their cost of service. Conversely, Alberta's electricity sector operates on a competitive market basis.

This study concentrates on examining how coordinated Distributed Energy Resources (DERs) at the distribution grid level interact with wholesale market-based power system operations. In the United States, FERC Order 2222 has facilitated the participation of DERs alongside traditional resources in regionally organized wholesale markets through aggregations. In competitive markets, price is influenced by the dynamics of supply and demand. The supply aspect of the wholesale electricity market includes both generation and transmission, which must be sufficient to meet the simultaneous, instantaneous, and reliable demand of all end-users.

Key supply factors such as fuel costs, capital costs, transmission capacity and its constraints, and the operational characteristics of power plants influence pricing. Demand-side changes also play a role. In real-time operations, demand is constantly fluctuating. The integration of DERs into this mix adds further uncertainty to electricity demand. To address these factors, ISOs have implemented Locational Marginal Pricing (LMP).

Locational Marginal Pricing

Locational Marginal Pricing (LMP) is a method used in wholesale electric energy markets to reflect the value of electricity at various locations, taking into account load patterns, generation, and the physical limitations of the transmission system. RTO/ISO markets determine an LMP for each point on the power grid, representing the marginal cost to supply load at that specific location, considering the set of generators in use and transmission system constraints. LMP comprises three components: an energy charge, a congestion charge, and a charge for losses in the transmission system. In most ISOs, LMPs are recalculated every five minutes. The LMP for a load zone is the weighted average of the LMPs at all nodes within that zone.

In an ideal scenario with no transmission constraints or losses, all LMPs would be identical, reflecting only the cost to supply the next increment of load. The lowest-cost energy offer would meet this additional megawatt of load, and electricity from this generator could flow to any point on the transmission system. However, LMPs typically vary across locations due to transmission and reserve constraints that prevent the next-lowest-cost megawatt of electricity from reaching all grid locations. Additionally, even when the lowest-cost megawatt can reach all areas, the marginal cost of physical losses in the system leads to varying LMPs.

"Price-Taker" versus "Price-Maker" Nodes in Wholesale Markets

Generation units and Load Serving Entities (LSEs) that participate in the market for selling and buying energy are known as price-maker nodes. These entities influence the market clearing price determined by the ISO/RTO through the intersection of supply and demand curves. Conversely, Distributed Energy Resources (DERs) located behind a customer's meter are unable to engage in the wholesale market, thus categorizing them as price-takers. A price-taker in the power systems is a participant who lacks the ability to influence market prices and is often referred to as a "behind-the-meter" resource. The constraints faced by DERs are further explored in subsequent sections.

The electricity bill for end-users, who are typically considered "price-taker" nodes, comprises several components that contribute to the total cost. These components vary across different regions and operators. The three primary elements that make up the electricity price are: i) the cost of generating electricity; ii) the cost of transmitting electricity across long distances via high-voltage transmission networks; and iii) the cost of distributing electricity through low-voltage distribution networks by retailers.

Demand Response

Electricity demand tends to be unresponsive to price fluctuations, meaning it does not significantly decrease when prices increase. This lack of sensitivity is partly because most end use electricity consumers do not experience real-time electricity pricing. To address this, some utilities and grid operators have implemented methods to encourage consumer participation in managing demand. Demand response (DR) involves a short-term, voluntary reduction in electricity use by end-users, typically activated by issues of grid reliability or high prices in the wholesale market. Customers who reduce (or sometimes just agree to reduce) their electricity usage during these periods receive compensation.

Various demand response programs are managed by a range of entities, such as market operators, utilities, load-serving entities, and third-party aggregators, both at the retail and wholesale levels. In this study, our emphasis is on the non-dispatchable form of demand response. This focus has been chosen because the main goal is to enable prosumers to independently make decisions about their energy usage and trading, within the framework of a peer-to-peer market arrangement, while maintaining a connection to the central grid.

NOTES

1. In models of power systems at the transmission level, loads are typically treated collectively. That is, loads are depicted in a consolidated fashion. As a result, the detailed granularity of load modeling is not captured in large-scale grid models. Consequently, operators lack insight into small-scale DERs, even though these devices are equipped with advanced sensors and controllers.

2. Established pursuant to Federal Energy Regulatory Commission (FERC) Order No. 2000, an RTO is a federally regulated independent entity that is responsible for: i) Managing, providing access to, and monitoring transmission facilities under its control; ii) Maintaining grid stability and reliability; iii) Matching electricity demand to supply; iv) Facilitating competition among wholesale electricity suppliers.

3. An independent and federally regulated entity that coordinates regional transmission to ensure non-discriminatory access to the electric grid and a reliable electricity system. ISOs were formed following the issuance of FERC Orders Nos. 888 and 889 by the Federal Energy Regulatory Commission (FERC) to encourage utilities to work together to ensure they provided non-discriminatory access to their transmission lines to non-utility-owned power plants. ISOs are not the only transmission operators in the US. There are also regional transmission organizations (RTOs) that perform generally the same functions as ISOs, but RTOs have jurisdiction over a larger geographic area.

4. The energy *trades*, represented by the dotted lines in the diagram, are different from physical energy *flows*. These flows are dealt with, primarily in the physical sphere, while the market sphere manages the trades.

BIBLIOGRAPHY

AlAshery, Mohamed Kareem, Zhehan Yi, Di Shi, Xiao Lu, Chunlei Xu, Zhiwei Wang, and Wei Qiao. 2020. "A Blockchain-Enabled Multi-Settlement Quasi-Ideal Peer-to-Peer Trading Framework." *IEEE Transactions on Smart Grid* 12 (1): 885–896. doi:10.1109/TSG.2020.3022601.

Andoni, Merlinda, Valentin Robu, David Flynn, Simone Abram, Dale Geach, David Jenkins, Peter McCallum, and Andrew Peacock. 2019. "Blockchain Technology in the EnergySector: A Systematic Review of Challenges and Opportunities." *Renewable and Sustainable Energy Reviews* 100: 143–174. doi:https://doi.org/10.1016/j.rser.2018.10.014.

Bade, Gavin. 2018. "MISO States Push for Authority over DER in Electricity Markets at FERC Meeting." https://www.utilitydive.com/news/miso-states-push-for-authority-over-der-in-electricity-markets-at-ferc-meet/521089/.

Cali, Umit, and Ozan Çakir. 2019. "Energy Policy Instruments for Distributed Ledger Technology Empowered Peer-to-Peer Local Energy Markets." *IEEE Access* 7: 82888–82900. doi:10.1109/ACCESS.2019.2923906.

Dimeas, A.L., and N.D. Hatziargyriou. 2005. "Operation of a Multiagent System for Microgrid Control." *IEEE Transactions on Power Systems* 20 (3): 1447–1455. doi:10.1109/TPWRS.2005.852060.

Directorate-General for Energy. 2015. *Study on the Effective Integration of Distributed Energy Resources for Providing Flexibility to the Electricity System.* European Commission. doi:https://energy.ec.europa.eu/publications/study-effective-integration-distributed-energy-resources-providing-flexibility-electricity-system_en.

Doan, Hien Thanh, Jeongho Cho, and Daehee Kim. 2021. "Peer-to-Peer Energy Trading in Smart Grid Through Blockchain: A Double Auction-Based Game Theoretic Approach." *IEEE Access* 9: 49206–49218. doi:10.1109/ACCESS.2021.3068730.

Federal Energy Regulatory Commission (FERC). 2020. "FERC Order No. 2222: Fact Sheet." https://www.ferc.gov/media/ferc-order-no-2222-fact-sheet.

Han, Dong, Chengzhenghao Zhang, Jian Ping, and Zheng. Yan. 2020. "Smart Contract Architecture for Decentralized Energy Trading and Management Based on Blockchains." *Energy* 199. doi:https://doi.org/10.1016/j.energy.2020.117417.

Justia US Law. n.d. "National Association of Regulatory Utility Commissioners v. Federal Energy Regulatory Commission, No. 19–1142 (D.C. Cir. 2020)." https://law.justia.com/cases/federal/appellate-courts/cadc/19-1142/19-1142-2020-07-10.html.

Matheson, Jim. 2019. "Will FERC Trample State and Local Authorities in DER Rulemaking?" https://www.utilitydive.com/news/will-ferc-trample- state-and-local-authorities-in-der-rulemaking/552069/.

Mengelkamp, Esther, Johannes Gärttner, Kerstin Rock, Scott Kessler, Lawrence Orsini, and Christof.Weinhardt. 2018. "Designing Microgrid Energy Markets: A Case Study: The Brooklyn Microgrid." *Applied Energy* 210: 870–880. doi:https://doi.org/10.1016/j.apenergy.2017.06.054.

Mishra, Sakshi, Roohallah Khatami, and Yu Christine Chen. 2023. "Towards Harmonious Decentralization of Energy Systems: A Vision of Interoperable Peer-to-Peer Energy Markets." arXiv.org.

Morstyn, Thomas, and Malcolm D. McCulloch. 2019. "Multiclass Energy Management for Peer-to-Peer Energy Trading Driven by Prosumer Preferences." *IEEE Transactions on Power Systems* 34 (5): 4005–4014. doi:10.1109/TPWRS.2018.2834472.

Morstyn, Thomas, Niall Farrell, Sarah J. Darby, and Malcolm D McCulloch. 2018. "Using Peer-to-Peer Energy-Trading Platforms to Incentivize Prosumers to Form Federated Power Plants." *Nature Energy* 3: 94–101. doi:https://doi.org/10.1038/s41560-017-0075-y.

Pudjianto, D., C. Ramsay, and G. Strbac. 2007. "Virtual Power Plant and System Integration of Distributed Energy Resources." *IET Renewable Power Generation* 1 (1): 10–16. doi:10.1049/iet-rpg:20060023.

Seven, Serkan, Gang Yao, Ahmet Soran, Ahmet Onen, and S. M. Muyeen. 2020. "Peer-to-Peer Energy Trading in Virtual Power Plant Based on Blockchain Smart Contracts." *IEEE Access* 8: 175713–175726. doi:10.1109/ACCESS.2020.3026180.

Statisca. n.d. "Revenue of the Electricity Industry in the United States in Selected Years from 1970 to 2022." https://www.statista.com/statistics/190548/revenue-of-the-us-electric-power-industry-since-1970/.

Thomas, Lee, Yue Zhou, Chao Long, Jianzhong Wu, and Nick Jenkins. 2019. "A General Form of Smart Contract for Decentralized Energy Systems Management." *Nature Energy* 4: 140–149. doi:https://doi.org/10.1038/s41560-018-0317-7.

Ton, Dan T., and Merrill A. Smith. 2018. "The U.S. Department of Energy's Microgrid Initiative." *The Electricity Journal* (Elsevier) 25 (8): 84–94. doi:https://doi.org/10.1016/j.tej.2012.09.013.

[AQ8]UNDP. 1999. *Decentralization: A Sampling of Definition.* Joint UNDP-Government of Germany Evaluation of the UNDP Role in Decentralization and Local Governance. http://web.undp.org/evaluation/documents/decentralization_working_report.pdf.

United National Climate Change. n.d. *The Paris Agreement.* https://unfccc.int/process-and-meetings/the-paris-agreement.

Wang, Shen, Ahmad F. Taha, Jianhui Wang, Karla Kvaternik, and Adam Hahn. 2019. "Energy Crowdsourcing and Peer-to-Peer Energy Trading in Blockchain-Enabled Smart Grids." *IEEE Transactions on Systems, Man and Cybernetics* 49 (8): 1612–1623. doi:10.1109/TSMC.2019.2916565.

Wolf, Gene. 2020. "The Reality of Virtual Power Plants." https://www.tdworld.com/distributed-energy-resources/article/21122679/the-reality-of-virtual-power-plants.

Chapter Seven

The Question of Sustainable Energy in the Global South

Intertwining Investments Challenges, Innovation Questions and Policy Pathways

Gz. MeeNilankco Theiventhran

Energy transitions are unfolding in the global South. The global call for de-carbonising energy systems has made renewable energy a buzzword. Renewables headline the green, clean discourses and fit well with the government's environmental ambitions, green private sector initiatives, and people's quest for energy security. Over the past decade, renewable energy has become a central part of policy discussions in the global South owing to increasing energy demand and environmental commitments (Kumar et al., 2021). Transitions based on renewables have traditionally focused heavily on technological fixes and innovative advancements (Köhler et al., 2019; Sovacool et al., 2020). On the one hand, innovation has opened new pathways for the diffusion of renewables. However, on the other, it has created justice and equity issues, especially not only in the global South. This has highlighted the need to understand energy transitions within a complex socio-technical framework.

A successful and socially positive transition to a low-carbon society requires understanding community practices, technologies, infrastructures and institutions and the carbon impacts these. Moreover, it requires an understanding of the links between these components, their dynamics, and the options and potential 'levers' involved. However, much of the current energy research addresses technical solutions and innovations that are easy or cheap to mass-produce and achieve "quick fix" emission reductions (Cherp et al., 2018; Sarren & Haarstad, 2018; Meadowcroft, 2009). Whilst there is also interest in issues relating to energy security, issues of equity and justice have remained peripheral to energy research (Williams & Doyon, 2018; Mundaca et al., 2018). Exploring the ethics of energy systems and the power relations they embody, including procedural justice issues, is essential to advancing research around energy democracy (Burke & Stephens, 2018).

Through a Sri Lankan case study, this article illustrates how one renewable energy project—rooftop solar—can undermine energy justice and deprive the underprivileged of the benefits of renewable energy. It demonstrates that even when renewable energy projects are introduced and implemented to benefit the underprivileged, they may only benefit the privileged. Systemic weaknesses, policy obstacles and profit-based motivation can undermine broad social participation in renewable energy projects, even though the hope has been that innovation and technology will address the justice and equity questions.

Sri Lanka is a low-middle-income country with ever-increasing energy demand and growing resource consumption, significantly impacting the environment (ADB & UNDP, 2017; ADB, 2015). Sri Lanka's geography and climate favour solar, and almost no fossil fuels have been found. Over the years, Sri Lanka has relied almost entirely on coal and oil imports to produce electricity, draining the country's foreign exchange reserves (World Bank, 2019). Sri Lanka's international climate commitments, coupled with increasing energy demand, have made Sri Lanka's transition to renewables a very high priority. However, whilst political manoeuvring, policy complications and vested interests have hindered this transition, the top-down approach towards energy transition has undermined community involvement and left behind a section of the Sri Lankan society who are vulnerable but could contribute to making the transition more equitable.

During the past two decades, local communities have received increasing attention as locations for generating and owning renewable energy projects (Walker et al., 2007). In energy transitions, however, the community has many meanings and facets relating to its scale, location, networks, processes and identities (Walker, 2011). Communities need to be central to any renewable project, and failing to ensure this may create equity issues and risk leaving local populations behind.

Whilst the concept of energy transition often refers mainly to the energy systems themselves, it must be understood in the broader sense, which includes both energy efficiency and energy reductions through changes in industrial products, lifestyle changes, and other shifts in consumption. The goal of "decarbonisation" similarly refers to much more than eliminating fossil fuels (Kern & Rogge, 2016; Calver & Simcock, 2021). In this chapter, we focus on a critical component of the global energy transition, namely the transition to renewable energy sources—in our case, solar. However, the focus is not on renewable energy as such but on social and community issues and implications, i.e., social sustainability aspects, of the transition.

MATERIALS AND METHODS

Clean energy transition literature and policy focus on technological solutions and innovation (Köhler et al., 2019; Geels & Kemp, 2007; Turnheim & Geels, 2013). In recent academic literature, much attention has been devoted to the role of innovations in energy transitions. Innovations act in multiple ways as 'push' and 'pull' factors; technological advances, improved efficiency, and cost reduction have made renewables increasingly competitive. There is, however, a concern that the renewable energy sector has yet to make emerging new clean energy technologies, such as solar and wind, accessible to low-income communities, both rural and urban. Clean energy companies prioritize marketing their products to customers financially capable of adopting them. Those less well-off still await their turn, as they have done for decades.

The Multi-Level Perspective (MLP) approach to socio-technical transitions has developed into an influential strand of literature about transitions. Combining concepts from evolutionary economics, science and technology studies, structuration theory and neo-institutional theory, MLP builds on a framework of niches, regimes and landscapes as three functionally distinct but interrelated levels that shape the process of socio-technical transitions (Geels, 2002; Geels and Schot 2007, 2010). MLP is one of the frameworks developed to understand how the transition occurs within socio-technical systems to shift to more sustainable conditions (Markard et al., 2012). The MLP framework describes three levels in transition processes: the first is the niche level, the initially small and unrecognised area where innovations take place and new technologies emerge; second is the regime level, where existing institutions and governance structures are situated; and the third is the landscape level, where various external actors and pressures that affect transition processes are situated. These levels are not hierarchical but interrelated entities facilitating or hindering systems from following a particular development path. On each level, various decision-making bodies and persons interact, form coalitions or conflict with each other to achieve the goals of their respective agency structures (Grin, 2008; Kemp et al., 2007).

MLP is based on the assumption that stable regime conditions exist. Most regimes could be more stable in the global South, especially in post-conflict transitional states. MLP stresses that niche-regime-landscape processes should be aligned for a transition (Geels, 2002). In transitional societies in the global South, transitioning is happening, but whether the niche-regime-landscape processes are aligned needs investigating. MLP has come under criticism from two directions. First, scholars argue that MLP as a framework fails to note the problems of politics and agency in transitions (Medowcroft, 2011; Grin et al., 2011). Second, the applicability of MLP to a global South context needs to be clarified since it was built upon experiences and notions from the West, and the theorization is done through the Western philosophical tradition. Most MLP research has focused on developed countries, and more research needs to be done on applying the MLP framework to understand energy transitions in the global South. This is where the post-colonial critique becomes useful.

Post-colonial thinkers argue that Western-centric development and technological approaches create problems with equity and justice. Post-colonial perspectives call for a historical re-reading of energy challenges in light of the diverse colonial experiences of different countries and regions (Baptisa, 2008). The post-colonial critique points towards a global cultural legacy of oppression and exclusion (Bridge, 2009; Anderson 2002). It seeks to deconstruct Western understanding and notions and challenges the dominant ways of framing our social and political understanding of societies. It focuses on unequal and unsustainable local and global relations of power (Bryant, 1998; Mbembe, 2012). Post-colonial thinkers argue that we must break with the narratives of 'development' and 'modernisation' that support an idea of Western modernity, or Occidentalism (Chandhoke, 1991; Kothari, 2002; Venn, 2000), the model of social advancement. Therefore, post-colonial critique argues the need to look at local realities through local lenses rather than preconceived Western notions. These critiques have much validity and indicate a need to reconsider our understanding of energy transition and our approach to achieving it when addressing developing country contexts.

This raises the question of whether MLP is an adequate framework to understand energy transitions in the global South, especially in a post-war context. Academic literature on MLP focuses on the 'niches,' hence technological innovations and solutions. However, in the case of the global South, the 'regimes' and 'landscapes' may be far more complex and varied than in more stable Western contexts and tend towards different outcomes. This may apply particularly in post-war settings where regions are more widely disparate and advantaged or disadvantaged.

This study uses a case-study research design that allows for an in-depth explorative approach to understanding the rooftop solar development in Sri Lanka and the associated spread of solar PV. The study applies process tracing to explore the systematic or generalisable pathways towards renewable energy diffusion. Data collection was done mainly through fieldwork, and secondary data was based on extensive document studies of policies, regulatory procedures, research literature, news articles, and other secondary sources. The fieldwork was conducted in three phases. The first phase was done in November and December 2018, where potential rooftop solar customers, companies that provide solar rooftops and policymakers were interviewed. The second phase of the fieldwork was conducted one year later, from November 2019 to January 2020, and the same group of people were interviewed. The final phase of the fieldwork was scheduled for October 2020, but due to the COVID-19 pandemic became impossible and was therefore instead conducted virtually, with reasonable success. In phase one, 16 interviews and two focus group discussions were held. In the second phase, 19 interviews and three focus group discussions were held. During the virtual fieldwork, the data was collected after communicating with 22 persons through various communication platforms, including text messages, emails, phone conversations, Skype and Zoom.

ROOFTOP SOLAR IN SRI LANKA: FROM CHEQUERED HISTORY TO UNCHARTED TERRITORY

The Stockholm Conference on the Environment in 1972 and its focus on renewable energy provided an initial impetus for the Sri Lankan Government to study alternatives (Zubair, 2001). Nuclear power was an area of interest at the time. However, others, such as solar photovoltaics and wind power, were examined to complement the existing hydro and thermal power generation, with Solar Home Systems (SHS) in particular being earmarked as a poverty reduction strategy.

The energy crisis in 1973 further prompted the world to look towards alternative and renewable energy technologies; the Sri Lankan Planning Ministry appointed a committee to study the area. In 1975 the Ministry of Finance & Planning established an Energy Planning Cell, which continued to keep abreast of PV technology development. The high cost of solar photovoltaics was seen to be a barrier to its immediate application in Sri Lanka, even though the costs were expected to reduce drastically (Perera, 1992).

In 1975 the United Nations Environmental Programme (UNEP) aimed to implement energy pilot projects in the developing world. UNEP decided to locate demonstration projects in Latin America, Africa, and Asia. Given Sri Lanka's established interest in renewable energy, UNEP and the Sri Lankan government decided to situate the 'Rural Energy Demonstration Centre for the Asian Region' in the southern part of the country. The work began in 1977 in the village of Pattiyapola (Gunaratne, 1992).

The US$200,000 project included a 10kW solar PV array, 20kW of wind power, a 300kWh per day battery bank and a 50kW biogas generator. These provided electric power for lighting and a water supply for 200 inhabitants of Pattiyapola village. The combined system could generate 50,000 kWh of energy per month at 230V AC and was the first system of solar photovoltaics to be installed in Sri Lanka. The PV modules were produced by the Solar Electric Corporation of the USA (Fernando, 1991; Williams, 1991).

The project was commissioned in 1979 but failed after a few years of operation because the technologically sophisticated inverters and system controllers became obsolete quickly, and replacement parts were not available. Also, the Ceylon Electricity Board (CEB), the government authority responsible for generating electricity in Sri Lanka, and managing the project, needed to allocate more funds to maintain the systems (Williams 1991; Gunaratne 1992). However, the project manager kept part of the system operational using locally available resources. Thus, only the biogas generator operated for a few hours to provide some power to the village. Even though the solar array produced electricity, the batteries and the inverters were not functioning. The project site was finally closed down in 1990. The project's failure had a significant social impact, where people started losing trust in solar and saw new technologies as disabling their upliftment and a way of leaving them behind. It is important to note that the program was initiated with interest, and Sri Lankans were looking towards alternatives.[1]

In 1980, the CEB established the Energy Unit to promote PV and other renewable technologies. The idea was to stimulate the market until private sector organizations began to promote these commercially. The CEB Energy Unit (CEBEU) undertook to supply and install systems using imported PV modules and other locally developed systems from its head office in the capital Colombo (Perera, 1992). Periodic newspaper advertisements attracted people from rural areas to travel to Colombo and purchase systems. The units were sold at approximately US$350–500, and about 600 domestic systems were sold between 1983 and 1989 (Williams, 1991). Over 150 systems using 8.5, 32 and 44W PV modules were sold in 1983/4, and from 1985 to 1989, the CEBEU sold 450 systems with PV modules of 16.5, 20, 33 and 44 W. A typical system included the PV module battery, 12V lamps and the cable; no charge controllers were included. The user did the installation after receiving instructions from the CEBEU staff (Gunaratne, 1992). The CEBEU has reported that even though there was general satisfaction amongst users, frustrations stemmed from the short life of the batteries (2 years) and the 12V lamp circuits (6 months) (Williams, 1991).

In 1985, the CEBEU established a solar-powered central battery charging station in a southern province village called Bundala. Ten 44W PV modules served 52 homes using automobile batteries to power a few 12V DC lamps and DC power televisions. A trained technician from the village was employed to charge batteries and collect money. In 1987, a similar centre was established in a second southern village called Kolaweheramankade. Here, 32 33W PV modules served a community of 100 homes (Fernanado, 1991). Both these stations have been abandoned since the period of civil unrest that plagued the country in 1988/89, affecting the rural areas the most.

In the 1990s and early 2000s, there needed to be more interest from the Sri Lankan government. However, Sri Lanka was keen on electrification, extending the national grid and connecting the islands through the national grid. The civil war began in the 1980s and became more aggressive conventional warfare. In the 1990s, the government primarily sought safe energy options due to the threat of attacks on energy supplies. In 2001, there was rapid growth in the SHS market due mainly to financial incentives provided by the World Bank through the Energy Services Delivery (ESD) Project (and its successor, the Project on Renewable Energy for Rural Economic Development) and by some provincial governments.[2] The aggressive marketing initiatives of several competitors in the SHS market also strengthened the efforts to expand the SHS market. This was further aided by innovative easy payment schemes introduced by micro-financing institutions such as Sarvodaya Economic Enterprise Development Service (SEEDS). The total SHS sales in Sri Lanka increased from about 1000 by June 2000 to over 15000 by February 2002. This was estimated to have increased to about 25,000 by 2003 (Martinot et al., 2001). This was thus a period of relative success.

Households and businesses without access to central electric, gas or fuel infrastructure often characterize remote rural areas of developing and emerging countries. Despite their relatively low energy demand, due to long transport distances and intermediate trade, the costs for meeting daily energy needs in these areas are proportionally higher than in areas with access to central energy infrastructure (Modi et al., 2005). Poor rural people—who must spend a large share of their income and time acquiring fuels for inefficient lighting and cooking purposes—must be specially targeted by programs that provide affordable, clean and modern energy services (Pachauri & Spreng, 2004). The widespread expectation is that this population could save significant amounts of fuel, money and time given more efficient energy solutions (ESMAP, 2003). The United Nations points out that access to energy services will have multiple positive effects, such as increasing productivity in the agricultural, industrial and tertiary sectors (ESMAP, 2003; UNDP, 2007).

In the mid-1990s, a national program established by the government of Sri Lanka with the World Bank and the Global Environment Facility (GEF) was coordinating electrification projects in rural areas of Sri Lanka. The World Bank, which ran the programme, argued that 'modern energy services directly contribute to economic growth and poverty reduction because new opportunities for income generation at the household level are created, unit costs can be reduced, and additional revenues from agriculture or animal husbandry can be achieved by permitting pumping for supplementary irrigation. Besides these direct impacts, indirect contributions to economic growth are expected as time can be saved, and resources can be used for other productive activities. Other positive impacts of modern energy services are improved health and education and reduced local environmental degradation (World Bank, 2002).

Ramani and Heijndermans (2003) differentiate the energy needs of the poor into "basic need applications," such as cooking, heating, lighting and comfort in households or communities, and "productive needs applications" in agriculture, home-based micro-enterprises and rural industries and services. They suggest providing microcredits "to overcome the poor's lack of purchasing power." The above illustrates how solar policy in Sri Lanka included considerations of poverty reduction and energy justice aimed at the neediest groups.

CASE STUDY: SOORYA BALA SANGRAMAYA (SBS) PROJECT

In 2016, the Sri Lankan government launched a community-based power generation project titled *Soorya Bala Sangramaya* (Battle for Solar Energy) to promote the setting up of small solar power plants on the rooftops of households and religious places, hotels, commercial establishments and industries. It was expected to add 200 MW of solar electricity to the national grid by 2020 and 1000 MW by 2025 (ADB and World Bank 2017). Under this program, customers can generate and use electricity on their premises. They can sell any excess electricity to the national grid or bank it for later use. The customer can select a preferred option from the following three schemes: Net Metering, Net Accounting, and Net Plus, according to the electricity usage.[3]

The aim was to introduce 1,000,000 Rooftop Solar PV units as a national program during the ensuing ten years. The objective is for at least 20% of the electricity customer base to become electricity producers. In 2017, Sri Lanka's Ministry of Finance announced that the Asian Development Bank had provided US$50 million to initiate a loan scheme to help homeowners and enterprises install rooftop PV systems under the 'Battle for Solar Energy' program. The project allows customers to get a loan from banks to purchase rooftop solar PV. The credit line offered by the ADB is given to the Ministry of Finance. The ministry manages the funds and channels them through selected Banks known as Participating Financial Institutions (PFI).

Despite the plans and policies, and financial assistance from the ADB, the rooftop solar PV project has yet to produce the expected results. Most rooftop solar PV projects under the program are implemented in urban centres, mainly in Colombo and its suburbs. Only a few of the nine provinces have implemented the program. This project revolves around three key agencies: the bank offering the loan for the rooftop solar PV system, the private company which plans and provides the rooftop solar PV system, and the Ceylon Electricity Board, which approves the project. From start to finish, it is a 5-step process (Figure 7.1), wherein there are barriers and obstacles at every step.

Service Provider

The first and foremost player in the process is the service provider. This solar company has the approval and capacity to provide the customer with a rooftop solar PV system. On its website, Sri Lanka Sustainable Energy Authority has the approved list of registered companies in this respect. As of June 2021, the list comprises 380 companies. With the company's assistance, the customer must prepare a technical and financial proposal to submit to the CEB for approval initially and the bank for the loan.

The customer's critical challenge has been finding a suitable service provider, and customers gave different views on their options. Several interviewees pointed out that the service provider's main criteria are the customer's financial ability. Some point out that these service providers were unwilling to go the extra mile to facilitate a possible solar PV system. Even though technical and commercial frameworks exist, including technical guidelines and standards for solar rooftop systems, the service providers differ in their approach. There is a significant price difference between the solar PVs offered by the service providers, some are very expensive, but the logic is that they are of high quality. In contrast, some are comparatively cheap, catering to customers with financial restrictions. The lack of regulations has been a concern for the industry. Sustainable energy authorities tasked with regulating the companies need quality-checking processes, which has allowed low-quality products into the market. The following quote from a customer sums up the challenges the customers face in finding a service provider:

> In the past few months, I have spoken to more than six companies who are providing rooftop solar solutions. Some say it is a must to go for high-quality

Figure 7.1. The Process Diagram for Obtaining a Rooftop Solar PV System.
Source: Created by the authors of this chapter.

panels for longevity; some say our financial possibilities decide the quality of the panels; some are ready to provide the panels at a very low cost. I am not a technical person; I am unable to decide. Another area of concern is how much capacity can I afford. Some prioritise the need of the household, some prioritise the possible amount I can afford, and some prioritise the available space of the rooftop. (Interview with a customer, 21 May 2020)

The commodification of energy transitions has been crucial in undermining justice and creating inequities. It is important to remember that even though these rooftop solar programmers are named community programmes, they are facilitated by market logic and directed by neoliberal economic policies that govern Sri Lanka. The respondents continually pointed out that the utmost priority of the solar companies has been making maximum profit. So, these companies always welcome high-end customers. In several instances, companies who have facilitated the request of not-so-high-end customers have provided low-quality rooftop systems, which started giving trouble for the customers from the second year of installation.

Service providers acknowledged that this programme is structured in a way that is unfavourable for the people at the bottom of the economic pyramid. They argue that this programme fits well with the people with the financial capacity, linkages with the bank and high-end electricity users. For an average Sri Lankan electricity user, this programme is not worth it and not feasible since the payback period will be longer, and they will have to wait to enjoy the return-on-investment surplus. Further, obtaining a loan and finalizing the payments takes time, and it burdens the service provider, so they prefer to assist high-end customers with no issues with cash flow. Several interviewees who wished to have rooftop solar were unable to do so due to the lack of interest of the service provider due to their economic logic. The government's failure to provide incentives and substitutes has deprived the poor people of accessing this programme.

CEB Clearance

The technical proposal needs to be submitted by the customer to CEB requesting individual connection approval. CEB grants approval based on two criteria. First, the network absorption capacity should be sufficient. Second, the submitted technical proposal should meet the technical guidelines of the CEB. Recently, the CEB still needs to give clearance for applications seeking permission to have rooftop solar PV systems. One service provider noted:

We have more than 100 applications pending with our regional CEB office. We have not received any clearance from CEB since February 2021. There is

> no standard procedure; different CEB offices have different systems in place. Recently we were told that the possible installed capacity is determined by the electricity usage of the past six months of the customer. It has also undermined the rooftop solar PV. (Interview with a service provider, 5 August 2021)

Several service providers have pointed out that CEB's resistance to rooftop solar has been the main reason for not seeing broader solar diffusion in Sri Lanka. The critical point reiterated by several CEB officials, for their part, is that Sri Lanka's grid connection system is not geared to absorb renewable energy into it to accommodate the power generation from rooftops. So, it has been a problem to provide clearance to allow new households to connect to the grid from their rooftop solar for electricity production. Let us look at the geographical distribution of the rooftop solar diffusion under this program (ref. Figure 7.2). It is clear that the urban centres with better distribution and absorption capacity can have connections. It creates distributional injustice where peripherics are undermined. Energy experts outlined during the interviews that over the years, knowing well that Sir Lanka is moving toward renewables, CEB failed to overcome transmission/distribution constraints to accommodate the renewable energy infrastructures. This is a fundamental energy democracy question. All the people who access electricity should have the right to give back the electricity they produce to the primary grid to make the energy transition green and clean. Nevertheless, it has not taken place. This raises important questions about electricity distribution and equity arising from renewable energy developments.

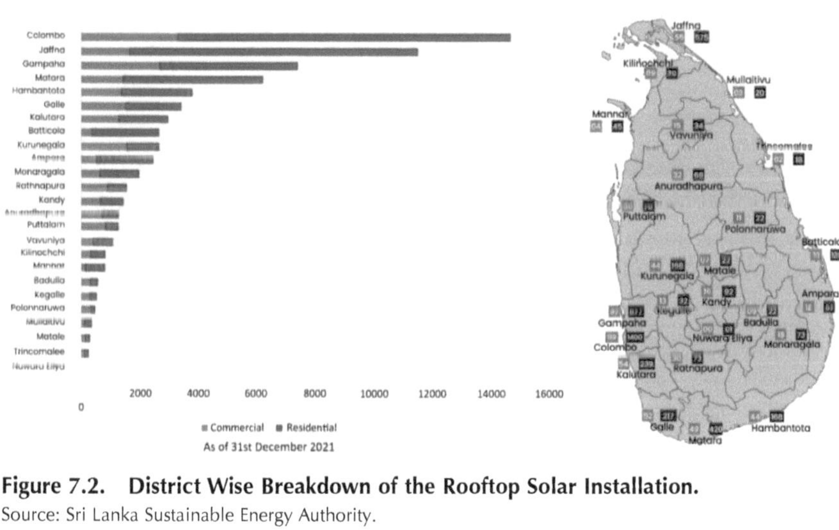

Figure 7.2. District Wise Breakdown of the Rooftop Solar Installation.
Source: Sri Lanka Sustainable Energy Authority.

The Loan from the Bank

The customer with the financial proposal then accesses the bank for a loan. The customer must submit the loan request, CEB approval, and technical and financial proposals. Upon assessing the customer's ability for repayment and proposals, the bank will approve the loan. In several instances, people have used their own money or different ways to fund their rooftop solar rather than seeking a bank loan through the ADB loan. In some instances, accessing an ADB loan at a low-interest rate has been extremely difficult. Once the bank agrees to provide the loan, the bank sends the central bank application to access the finance. Once they get clearance from the central bank, the bank approves the loan. This process takes a very long time. CEB's clearance certificate is usually time-specified; it allows the customer to install the solar PV within three to six months from the date of clearance approval. In several instances, the bank loan approval process has taken longer than the stipulated time frame allowed by the CEB. In recent months the ADB loan has been exhausted, so banks are not providing new loans due to the non-availability of funds from the Central bank. One banker outlined the process and struggles they encounter when processing the loan application:

> The first step in any loan application process is checking the creditworthiness of the customer. A customer who has never taken a loan is less creditworthy than a person who has taken a loan and paid it back. ADB loan is a cumbersome process; we need to send all the documents to the central bank for approval; as a state bank, we cannot provide any guarantee to the customer or the service provider. Whereas in private banks, in certain instances, they give a positive reply even before they get a green light from the central bank. They do it since they can take the risk, and alternatively, they will be able to facilitate a different loan. Overall other loans are easy to work with than this ADB facilitated loans. So generally, we do not promote this. (Interview with a Banker, 16 January 2020)

The banks focus on the creditworthiness of the customer. The process discriminates the disadvantaged people. Even though the rooftop solar PV system is profitable, the initial capital is high. It is not affordable for a customer with a modest income, and banks also see these customers negatively. This brings "just and equitable clean energy" into question—remembering that one of the critical drivers of this rooftop solar programme is to achieve the SDGs, where sustainable development goal 7 is affordable and clean energy for all.

Installation and Verification

Once the bank agrees to provide the loan, it releases the first payment for the service provider. The service provider is expected to install the rooftop solar PV system following the specifications in the technical proposal and the CEB approval. After the installation, an engineer from CEB, an accredited engineer registered with CEB, inspects the installed rooftop solar PV system to comply with project and utility guidelines and original technical proposal specifications. The accredited engineer issues a compliance certificate after inspection.

The quality difference in the solar panels, lack of skilled workers in the sector, and non-regularization of the service providers have meant that CEB's approval has often been delayed due to technical reasons. The essential criticism from the CEB technicians has been that the service providers fail to maintain the standards, which has led to delays. On the other hand, service providers argue that CEB unnecessarily delays the verification and does not follow set procedures. One energy expert pointed out:

> Several technicians are not trained enough to do the proper installation, which has become a problem. The reducing number of approvals also made solar companies diversify, intensified the problems. CEB complains that several companies do not meet the required technical standards, so the process has been difficult. (Interview, 14 February 2020)

One recurring theme during the interviews was the problem of technology diffusion. Even though there are more than 300 solar companies, most need more technical competence. Technological diffusion is taken for granted when these kinds of products are introduced. Notably, innovation literature must address justice and falsely assume that growth is the solution to justice (Papaioannou, 2011). Technology diffusion and innovation are intertwined with the economy in proportional terms. More financial capabilities mean more tech-savvy solutions, and the ones with minimalist purchasing power have to be satisfied with low technology, as the rooftop solar case suggests. This brings out a fair question of technology and innovation.

Grid Connection

With the compliance certificate, the customer must apply for a system connection with CEB. Once CEB signs the agreement with the customer, the agreement is submitted to the bank to release the remaining loan amount to the service provider. At this stage, there is rarely a delay. In this process, the service providers prefer customers who are not going for ADB loan facility since other loans or own financing is much easier and faster and ensures the customer's financial stability; in return, it is profitable for the service provider. This reinforces the inequities created in the process's previous steps.

Regional Disparities in the Rooftop Solar PV Diffusion

The challenges of inequity and justice also manifest geographically as regional disparities. The following diagrams outline the rooftop solar diffusion in Sri Lanka under the SBS program until 31 December 2021.

Figures 7.2 and 7.3 show how this program, intended to benefit the community at large, provided enduring inequality of outcomes at all spatial scales. The fieldwork data shows how the socio-spatial processes of the solar rooftop program reinforced existing social, political, and economic conditions and spatial injustice at the levels of the individual and the community. David Harvey (1973) stresses that capital markets and technological monopolies are the structural driving forces that characterize the capitalist mode of production and lead to an uneven distribution of resources in space and, ultimately, spatially uneven development. Urban centres have received the highest diffusion, and the peripheries have been left behind.

Even within the peripheries, there are disparities. An interesting finding is that the second largest rooftop solar diffusion occurred in the Northern province, which was affected by the civil war. Sri Lanka was involved in the three-decade civil war, which ended in 2009. The post-war setting in the Northern province incorporated the inequities created by the war. The rooftop solar programme has created new equity questions, as shown in Figure 7.4.

Notably, the Jaffna district, which has the most considerable solar diffusion, was entirely under government control. In contrast, the rest of the districts were partly or wholly controlled by the Liberation Tigers of Tamil Eelam (LTTE) and badly affected by war. Since the end of the war in 2009, the main goal has been electrification. Connection lines were installed for electricity distribution but not for a generation; however, in Jaffna, the transformers could take the feed-in electricity from rooftops. Reduced rooftop diffusion in the former war zones is due to three reasons:

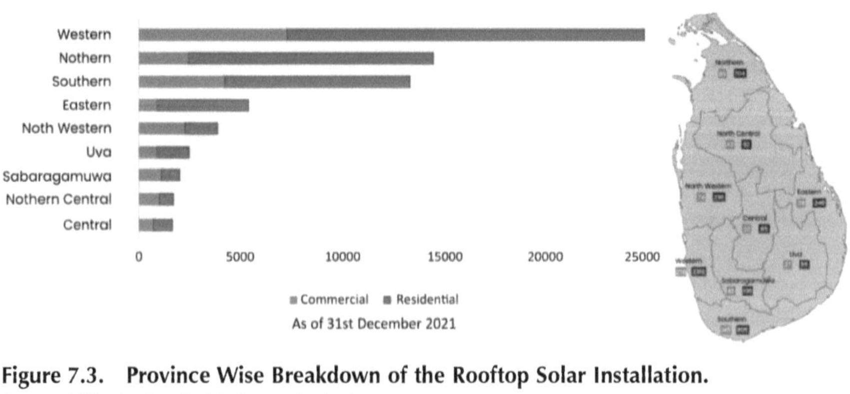

Figure 7.3. Province Wise Breakdown of the Rooftop Solar Installation.
Source: Sri Lanka Sustainable Energy Authority.

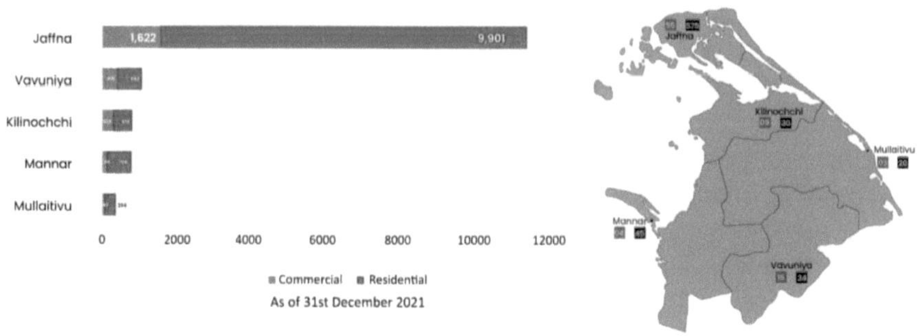

Figure 7.4. Breakdown of the Rooftop Solar Installation in the Northern Province.
Source: Sri Lanka Sustainable Energy Authority.

1. The people in the area have resettled again after losing everything due to civil war.
2. In war zones accessing the bank, loan facility has been complex.
3. The transformers are far from the possible rooftop solar PV houses, so it is difficult for CEB to give clearance to pursue rooftop solar due to technical reasons.

An interesting finding from the fieldwork in Jaffna is that in early 2020 the government of Sri Lanka reduced the bank interest rates for fixed deposits. This resulted in several persons in Jaffna deciding to take their savings and invest in rooftop solar, calculating that it would give them better value for money on a monthly basis rather than fixed deposits. This also contributed to increased interest in rooftop solar PV.

The Sri Lankan rooftop project experience is a typical example of top-down development projects in the global South, in this case, the rooftop solar programme. Today's mainstream economic thinking—often called neoliberalism—prioritises the market over people and the planet. In the Sri Lankan case, the banks' actions are understandable due to their way of operation. The criticism is directed at the neoliberal model, which undermines equity and calls for an increased role for the free market, privatization, reconfiguration of state welfare activities and reduced state interference in public affairs. As interviewers pointed out, these economic policies increase injustice and trigger socio-economic disparities, and the banks' actions concerning rooftop solar loans align with this. The neoliberal model excludes a section of the population from the economic opportunity to benefit from a new technology-driven process that has an overarching positive impact on society and the world.

Overall, obtaining rooftop solar and its associated challenges highlight three issues. First, the so-called "environmentally friendly" projects can create social inequalities. Therefore, instead of being exempted from verification procedures, they should be scrutinized to ensure that appropriate strategies are adopted through stricter procedures and, where possible, community participation in projects to guarantee that those who bear the brunt are not marginalized. Second, economic justice solutions should be at the forefront of renewable energy development projects to ensure equity and justice and no one is left behind. Third, there is a risk that innovations may not diffuse widely and only be available to some, thus creating opportunities for domination and exclusion.

The mechanisms used in the Sir Lankan rooftop program create access for some rather than others. The process acts as a barrier at every step of the way, as described above. The mechanism mainly structured on technology and finance creates new inequities and deepens or reconfigures existing ones.

The case study outlines the importance of agency in sustainability transitions, where the actors and institutions are embedded in transitions through the power structures. In the global South, regimes are fluid and socio-political implications impact the regime outcomes, whereas MLP considers regimes primarily as monolithic structures which actors will not influence. The post-colonial critique is valid since a local socio-political setting and political economy are always context specific. Viewing it through liberal political and economic lenses can misunderstand or fail to acknowledge the nuances and intricacies.

DISCUSSION: FOUR LESSONS FROM SRI LANKAN ROOFTOP EXPERIENCE

The preceding sections offered an overview of Sri Lanka's rooftop solar history and the challenges in making it inclusive and sustainable for all. The Sri Lankan case offers four broader lessons for renewable energy development in developing countries and the policymakers, practitioners and researchers.

THE ROLE OF AGENCY

The Sri Lankan case outlines the role of the agency in processes such as the diffusion of renewable energy. In the SBS rooftop program, it is clear that CEB, solar companies, and banks are the actors who decide the program's fate. The fieldwork data points out that harmonization and synchronization between these three institutions have yielded positive results and experiences; whereas difficulty caused by any one of these institutions has created uncertainties, this uncertainty then had a ripple effect on other agencies, and problems or failures resulted.

Figure 7.5 outlines the dialectic relationship between the institutions in the rooftop solar PV uptake in Sri Lanka and how it connects with policy, finance, and technology sectors in energy transitions. As shown in the case study, the interaction and dynamics within the agencies influence the planned renewable energy uptake outcome and chart the trajectories for sustainable and just energy transitions. Figure 7.5 displays how the findings from the case study can be generalized to understand energy transitions in the global South context.

The agency shapes how any new initiative is implemented; hence, the agency holds a vital place in the systems embodied in the energy transitions. Among agencies, there are always contestations over superiority and power, which is also an outcome of social unevenness. It is important to note that energy landscapes always have winners and losers, although their distribution can change over time (Bridge et al., 2018). The relationship among the agencies decides the patterns of social and geographical inequality and injustice and reflects asymmetries in the power, influence, and resources available to different actors.

Over time, institutions become entrenched and amass power and influence in their social-political landscape. This has been the case for all the institutions over time and their historical landscape in the structure of the state and society (Hall & Taylor, 1996). This historical landscape creates path dependence in the development of the institutions, and this path dependency is reinforced and reinvented through their relations with other institutions (Pierson, 2000). This creates structural lock-in, which positively or negatively impacts energy transitions.

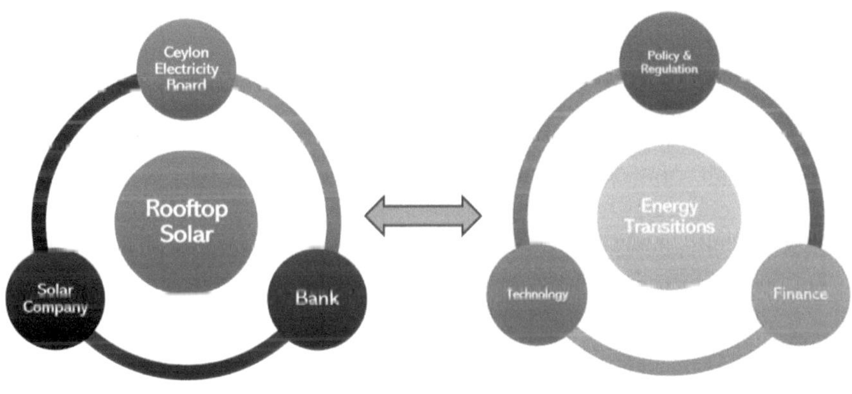

Figure 7.5. Interlinkages Between the Agencies.
Source: Created by the author of this chapter.

As shown in the Sri Lankan case, rooftop solar PV systems are subject to approval before connecting to the grid. From the point of view of a network operator and distributor, this makes technical sense, as every installed roof system changes the electrical load, the needs of its network, and the customer's energy demand profile. Therefore, every time a customer uses solar, the utility loses revenue. This challenge is particularly acute in emerging markets, where the most prominent companies and heavy electricity users are the first to opt for rooftop solar systems. When the largest energy consumers start to generate their electricity from their rooftop systems, their need to buy energy will decrease, and the utility will experience a loss. On the one hand, it is financially challenging, and on the other, it puts the utilities' inviolable monopoly under stress.

A key takeaway related to energy transitions is that the nature of the renewable energy design determines the agency's role. Most designs are based on "sustainable energy transitions" literature and theorization with global North experience. This prioritises innovative high-tech technology and market design as two critical parameters for transitions, and under this premise, the agency's role is structured. The case context of the global South presents different challenges and questions regarding agency, and non-understanding of the alternative configurations of energy transitions may cause issues of equity that undermine just and sustainable transitions.

THE ROLE OF POLICYMAKING

The history of Sri Lanka's rooftop solar initiatives and the two cases discussed in this article highlight the importance of clear policymaking and the ability to put those policies into practice. It has been a significant challenge for many developing countries where political fluidity, unstable financial situations and sectoral segregation of public affairs have dominated the political sphere nationally and locally. Policymakers often have little consensus on how to move forward, and the priorities vary with institutions. Even though all key agencies agree to achieve 'affordable, equitable and sustainable energy for all,' the understanding of how to achieve it has been different. It has led to inconsistent and competing policymaking on climate policies, energy planning and energy transition in the past few decades in Sri Lanka.

For any industry to prosper, it needs stable and secure regulations. It is vital to close the gaps between the government, the private sector, the public and the financial institutions. Policymaking is the necessary avenue to closing these gaps. To do so, policymaking needs to take a holistic approach. This problem is common to developing countries: but in most developing markets, regulations and policies such as those relating to energy are in their infancy or still developing. It takes time to design solutions that meet all stakeholders' needs, expectations and limitations. These may contradict each other—for example, where a largely coal-based power industry is faced with decarbonization or an energy company is mandated to promote energy savings or prosumers. Regulation—such as to ensure that any equipment on the market is technically adequate or to ensure consumer protection—is often lacking too.

Further, the target is to ensure affordable energy for the large low-income population segments and geographical access to rural areas outside the main centres. In that case, a bottom-up consultative approach is vital to avoid disempowerment, foster local skills and contributions, and address the equity question. Broad civic participation has been seldom in Sri Lanka—let alone embedded in law as in many of the rich countries. In the absence of this, those populations whose voice or agency is weakest, to begin with, risk becoming even further marginalized and left out of the energy transition with its potential social as well as environmental benefits.

Economic globalization has undermined the local economic and social needs and solutions. What is needed is more localization. Globalization and the programs prescribed by multilateral and bilateral donors have often created inequity, unsustainability, and unreliability. The story of Sri Lanka's solar projects underlines the need for localization, where reaching low-income people is more important than technological advancement and high-tech solutions. It is evident through the SBS case that the civil war also influenced solar localization, often avoiding and thus further marginalizing certain ethnic areas. Again, it confirms the post-colonial critique and the need for contextual and differentiated approaches.

THE ROLE OF BUSINESS AND FINANCE

The lack of funding is one of the critical reasons for the disappointing growth of solar rooftop systems in emerging countries and a transition to a fair, reliable and sustainable energy system. Nevertheless, to benefit the poor, funding must be targeted, accessible and as non-bureaucratic as possible. Access to finance through government incentives can foster the democratization of rooftop solar PV programs where people at the bottom of the financial pyramid will also be able to participate and contribute. Financial availability through low-interest loans from banks, government subsidies, and international grants will help more local participation in two ways: community participation and local private sector sustainability.

A stable local environment is imperative for community engagement in the renewable energy sector, such as upscaling rooftop solar PV systems. Financial viability is a crucial parameter for local involvement where the community feels they were considered and will participate in the PV program. The availability of funding will provide business for local private companies. The delays or stagnation in granting permission for PV systems in Sri Lanka has made several companies rethink their future. In turn, instability in the sector hinders the development of skills and quality in the industry.

In the Sri Lankan case, the lack of progress in the renewable energy sector can only be explained by the lack of funding or the complexity of implementation. One of the key obstacles to the industry's growth is that pricing and financing frameworks must be tailored to the local context—especially if a goal is to reach the poorest. For example, if we look at the value proposition of rooftop solar PV in any market, this is just one of many choices people might invest in. It applies to residential users as well as business owners. All have investment decisions to make. For business owners, this sometimes comes down to investing money in income-generating investments rather than cost-saving investments like rooftop solar PV. Moreover, we have noted that the costs were high for the poorest.

It is a problem worldwide, where the business model of energy producers often conflicts with energy efficiency goals or with self-generation by prosumer customers. There have often been poor outcomes when large energy utilities (who have the know-how and the technical control) are given a mandate for energy efficiency or on-site renewables. Government climate policies, business, and user interests collide unless the system is carefully crafted—adapted to local contexts—and efficiently applied.

THE ROLE OF TECHNOLOGY

Technology plays a crucial role in advancing energy transitions in industrialized countries. Theories and frameworks built around these successes have developed into assumptions about socio-technical transitions. Governance generally shifts towards more sustainable development (Grin et al., 2010). These theories emphasize technology as the pathway to sustainable transition goals. However, the cases discussed above note that overreliance on technology can undermine the sustainable transition. It is especially the case if technology development and choices do not include the non-technical, i.e., sociological and cultural, factors. The high-quality rooftop solar PV modules have been expensive and unaffordable except for middle-class customers, whilst cheaper low-quality PC modules have given constant problems for customers, creating a trust deficit among the local population.

Developing countries rely heavily on developed countries for technology diffusion, which creates dependency (Hansen et al., 2018; Köhler et al., 2019; Ramos-Mejía, Franco-Garcia, & Jauregui-Becker, 2018). The technology can also favour a neoliberal capitalistic model by which local communities' power is limited, their influencing capacity is reduced, and they become sidelined in many of the social and political aspects of the energy transition (Lachman, 2013). The Sri Lankan case shows that the contextual setting and dynamics influence technology diffusion and the impacts of technology. Understanding the socio-technical transitions with a Westernized theoretical approach may be to misunderstand the transitions in the global South. As discussed above, MLP is based on certain assumptions and can be a helpful tool to understand the transitions in the industrialised societies in the global North. However, for the global South, the settings are different. MLP, like other transition theories, is based on the liberal market model of delivery and a Western understanding of development with technocratic, top-down approaches. As discussed above, this does not fit for developing country context, and Sri Lanka's rooftop solar projects offer a typical case.

CONCLUSIONS

Transitions to renewable energy encompass multiple challenges at different levels. In a developing country context, a critical question is how to make these transitions just and inclusive. It demands the development of policies that apply more flexible tools, cooperation among institutions, and clearer government commitment. Dysfunctional politics and inbuilt conflicts between the institutions have undermined the progress in the energy transition towards achieving carbon neutrality which is envisioned by developing countries. The energy transition should also be stimulated and strongly influenced through public engagement and participation in policy, implementation and management. Inappropriate development paradigms, top-down perceptions about technological progress, and consistency mainly drive the policies.

There is a need to rethink local contexts, cultures and inequalities persisting in developing countries. Varied local realities, obstacles, and opportunities must take precedence over technological fixes and conventional business practices. In the global South, energy transitions at the local level should meet the local realities, especially in equity and market designs. High-tech innovations that are not sensitive to existing realities will likely create new inequities. There remains the complex question of how this can be avoided.

Energy transitions need to be configured to favour just and sustainable outcomes, with a particular view to improving the conditions of the poorest social groups and communities. It is necessary to counter or at least balance the neoliberal capitalistic mode of technological-financial fixes typically promoted by bilateral and multilateral agencies promoting projects like SBS in Sri Lanka and private sector investors. The experiences discussed above illustrate how projects intended for the benefit of marginalized communities can, on the contrary, benefit mainly those at a higher income level and, at worst, lead to further disempowerment and marginalization.

To summarise: the following are four key prerequisites for the achievement of successful as well as equitable energy transitions in the developing world. First is a localizing, participatory approach with a better understanding of varied local realities. Second, the consistent, inclusive policy built on improved inter-sectoral cooperation. Third, political willingness to move beyond outdated or colonial understandings of development. Fourth, financing models and mechanisms that disfavour the poorest need to be replaced with models that harness local skills, knowhow, business and management—in order to provide disfavoured communities not only with renewable energy but with the additional benefits and community improvements that are amongst the declared goals of the SDGs, and of solarization and similar programs in the field of sustainable development.

NOTES

1. Sri Lanka liberalized, and the open economy was introduced in 1977. This project, which was initiated in 1977 and started in 1979, was seen as one of the outcomes of the new economic model.

2. Sri Lanka enjoyed relative peace from 2001 to 2005 because Norway brokered a ceasefire between the Government of Sri Lanka and the Liberation Tigers of Tamil Eelam (LTTE). In this period the World Bank and international donors pumped vast amounts of money to improve the living of the people in Sri Lanka. They wanted a 'peace dividend,' which would help move the peace process in the right direction.

3. Net metering is a billing mechanism that credits solar energy system owners for the electricity they add to the grid. The consumer has to pay only for the net amount of electricity consumed during a month. If the solar electricity supplied to the grid is more than the electricity consumption during a month, the balance amount can be carried forward to future as a credit, up to a maximum of ten years. No payment will be made for the excess electricity supplied over the consumption under this scheme. In Net accounting, if the generated units of electricity using the solar panels fixed on houses/premises are greater than the amount consumed, the excess will be paid at the rate of LKR 22.00 per unit during the first seven years and from the eighth year onwards at the rate of LKR 15.50 per unit by the utility provider. Net plus involves getting paid for the amount of electricity generated using the solar panels fixed on the premises. Unlike the other two methods there is no linkage between the electricity consumption of the customer and the electricity generation. Two separate meters will be installed to measure import and export of electricity separately. The customer has to pay for the electricity consumed according to the existing tariff whilst the Utility Provider will pay for the total amount of electricity the customer generates.

BIBLIOGRAPHY

ADB (Asian Development Bank) and World Bank. 2017. *100% Electricity Generation through Renewable Energy by 2050: Assessment of Sri Lanka's Power Sector.* Available at: https://www.adb.org/publications/electricity-generation-renewable-energy-2050-sri-lanka (Accessed: 12 July 2020).

Anderson, Warwick. 2002. 'Introduction: Postcolonial Technoscience.' *Social Studies of Science 32*(5–6): 643–658.

Baptista, Idalina. 2018. 'Space and Energy Transitions in Sub-Saharan Africa: Understated Historical Connections.' *Energy Research & Social Science 36*: 30–35.

Bridge, Gavin.2018. 'The Map Is Not the Territory: A Sympathetic Critique of Energy Research's Spatial Turn.' *Energy Research & Social Science 36*: 11–20.

Bridge, G., Barr, S., Bouzarovski, S., Bradshaw, M., Brown, E., Bulkeley, H., and Walker, G. (2018). *Energy and Society: A Critical Perspective*. London: Routledge.

Bryant, L. R. 1998. 'Power, Knowledge and Ecology in the Third World: A Review.' *Progress in Physical Geography 22*: 79–94.

Chandhoke, N. 1991. 'Indigenous Social Science.' *Economic and Political Weekly 26*(21): 13–57.

Dharmadasa, Indrajith. 2016. *'Solar Energy Strategy for Sri Lanka: The Solar Village Solution for Sustainable Development and Poverty Reduction.'* In: Science and Technology for Society Forum, Sri Lanka, 7–10 September 2016.

ESMAP—Energy Sector Management Assistance Program. 2003. *Energy and Poverty: How Can Modern Energy Services Contribute to Poverty Reduction* Addis Ababa: Proceedings of a Multi-Sector Workshop.

Fernando, Cyril. 1991. *The Pansiyagama Project: Socio Economic Impact of Solar Electrification*. Colombo: Marga Institute.

Geels, Frank. 2002. 'Technological Transitions as Evolutionary Reconfiguration Processes: A Multi-Level Perspective and a Case-Study.' *Research Policy 31*(8–9): 1257–1274.

Geels, Frank, and Johan Schot. 2007. 'Typology of Sociotechnical Transition Pathways.' *Research Policy 36*(3): 399–417.

Geels, Frank, and Johan Schot. 2010. 'The Dynamics of Transitions: A Socio-Technical Perspective.' In John Grin, Jan Rotmans, and Johan Schot, eds., *Transitions to Sustainable Development*. London: Routledge, 11–104.

Grin, John. 2008. 'The Multilevel Perspective and Design of System Innovations.' In Jeroen van Bergh, and Frank Bruinsma, eds., *Managing the Transition to Renewable Energy. Theory and Practice from Local, Regional and Macro Perspectives.* Cheltenham: Edward Elgar, 47–79.

Grin, John, Jan Rotmans, and James Schot. 2011. 'On Patterns and Agency in Transition Dynamics: Some Key Insights from the KSI Programme.' *Environmental Innovation and Societal Transitions 1*(1): 76–81.

Gunaratne, L. 1992. 'Electricity for All by the Year 2000: Solar Electricity and Its Role.' Paper presented at Seminar on Electricity for All by the Year 2000, organized by the Association of Franco-Ceylonese Technologists, Colombo, Sri Lanka.

Hall, P. A., and R. C. R. Taylor. 1996. 'Political Science and the Three New Institutionalisms.' *Political Studies 44*(5): 936–957.

Kemp, Rene, Derk Loorbach, and Jan Rotmans. 2007. 'Transition Management as a Model for Managing Processes of Co-Evolution Towards Sustainable Development.' *International Journal of Sustainable Development & World Ecology 14*(1): 78–91.

Kothari, B. 2002. 'Theoretical Streams in Marginalised People's Knowledge(s): Systems, Asystems, and Subaltern Knowledge(s).' *Agriculture and Human Values 9*: 225–237.

Linden, David, and Thomas Reddy. 2002. *Handbook of Batteries*. New York: McGraw-Hill.

Markard, Jochen, Rob Raven, and Bernard Truffer. 2012. 'Sustainability Transitions: An Emerging Field of Research and its Prospects.' *Research Policy 41*(6): 955–967.

Martinot, Eric, Anil Cabraal, and Subodh Mathur. 2001. 'World Bank/GEF Solar Home Systems Projects: Experiences and Lessons Learned 1993–2000.' *Renewable and Sustainable Energy Review 5*(1): 39–57.

Mbembe, Achille (2001) *On the Postcolony*. Berkeley, CA: University of California Press.

Meadowcroft, James. 2011. 'Engaging with the Politics of Sustainability Transitions." *Environmental Innovation and Societal Transitions 1*(1): 70–75.

Pachauri, Shonali, and Daniel Spreng. 2004 'Energy Use and Energy Access in Relation to Poverty.' *Economic and Political Weekly 39*(3): 271–278.

Perera, Ruwan. 1992. *Energy Status of Sri Lanka: Issues, Policy, Suggestions*. Colombo: Institute of Policy Studies.

Pierson, Paul. 2000. 'Increasing Returns, Path Dependence, and the Study of Politics.' *The American Political Science Review 94*(2): 251–267.

PUCSL—Public Utilities Commission of Sri Lanka. 2005. 'Lessons from SHS in Sri Lanka Rural Areas.' Colombo: Public Utilities Commission of Sri Lanka.

Ramani, K. V., and Enno Heijndermans. 2003. *Energy, Poverty and Gender—A Synthesis*. Washington, DC: The World Bank.

Ratick, Samuel. 2005. *Solar Energy Adoption and Diffusion in Sri Lanka: An Analysis at Multiple Scales*. Worcester, MA: Clark University Press.

UNDP—United Nation Development Programme. 2007. *Energy and Poverty in Sri Lanka: Challenges and the Way Forward.* Bangkok: UNDP Regional Centre.

Venn, Couze. 2000. *Occidentalism: Modernity and Subjectivity*. London: Sage.

Walker, Gordon. *Environmental Justice: Concepts, Evidence and Politics*. London: Routledge, 2011.

Walker, G., S. Hunter, P. Devine-Wright, B. Evans, and H. Fay. 2007. "Harnessing Community Energies. Explaining and Evaluating Community-Based Localism in Renewable Energy Policy in the UK." *Global Environmental Politics 7*(2): 64–82.

Williams, Nigel. 1991 'Solar Serendipity: Photovoltaic Rural Electrification in Sri Lanka.' *Solar Today 5*(7): 14–22.

World Bank. 2002. 'Project Appraisal Document on a Proposed Grant to Sri Lanka for a Renewable Energy for Rural Economic Development Project.' South Asia Regional Office: Energy and Infrastructure Unit.

Zubair, Lareef. 2001. 'Challenges for Environmental Impact Assessment in Sri Lanka.' *Environmental Impact Assessment Review 21*: 469–478.

Chapter Eight

Heading Towards Net-Zero

India's Multilateral Partnerships to Promote Renewable Energy in the Post-Pandemic World

Aditi Basu and Elena V. Shabliy

India is one of rapidly growing economies that achieved its independence in 1947. Notably, it is the second most populous country in the world. India saw strong GDP growth in the third quarter of 2021, and Indian economy in 2022 has fully recovered the pre-pandemic real GDP level of 2019–2020 as stated by Union Minister of State for Finance Shri Pankaj Chaudhary.[1] The Minister also stated that "the revised growth for India is still higher than that of major advanced and emerging market economies."[2] For its rising population, India's demand for energy is rising. According to IEA, India is the world's third-largest energy consuming country, due to rising incomes, as well as improving standards of living. At the time of its independence, India heavily relied on coal for meeting its energy demands that led to more greenhouse gas emissions (GHG) emissions in the atmosphere. Since then, it has been looking towards alternative sources of energy in accordance with the Paris Agreement. To suffice its needs, the country has witnessed a transition to green and renewable energy, keeping in mind its sustainable development goals and net zero targets. Being the third largest producer of renewable energy (IEA), India heavily relies on solar, wind, green hydrogen and ammonia as prime sources of renewable energy.[3] From hydroelectricity, India's journey in green energy has opened multiple investment opportunities and multilateral partnerships for the world and the strategic importance of renewable energy in Indian foreign policy. From being a founder member of the International Solar Alliance to being the only country strictly adhering to the Paris Agreement, India has announced to achieve net-zero by 2070 and to meet 50 percent of its energy demands through clean energy sources by 2030 that is a significant step in the global fight against climate change. During the COVID-19 pandemic, there had been a slump in financial investments in renewable energy all over the world. In the post-pandemic world, India is expected to make rapid legislations and more multilateral partnerships in renewable energy as a middle power, linking the Global North and South. This chapter analyzes India's role as a leading Global South economy in the North-South politics of climate and energy negotiations focused on renewables. It focuses on India's role in bilateral and multilateral partnerships to combat the vulnerabilities of climate change and energy insecurity during COVID-19 and in the post-pandemic world, its role as a Middle Power in bridging the North-South divide in renewable energy partnerships towards attaining net-zero. The chapter presents a case study analysis of India's multilateralism in renewable energy, using an inductive method. It is qualitative research that involves a political and economic approach. Secondary data has been obtained through a thorough qualitative analysis of government reports, articles, books, research papers, journal chapters, websites and newspapers. Official reports uploaded to government sites have been studied. The methods allow a better understanding of the relationship between India and the globe in renewable energy diplomacy in the twenty-first century, in general, and in the post-pandemic world, in particular.

As a Global South nation and housing 20 percent of the global population, India has been affected by the catastrophic effects of climate change, given its geopolitical strategic location in South Asia.[4] Hence, climate being "a global common" drives both the Global North and South to act responsibly in order to mitigate the reasons of climate change that constitute the *global public good*. Also, the world has been experiencing energy crisis for the last decade, in general, and since COVID-19, in particular. This has led to the development of an international climate change-energy diplomacy paradigm that requires financing (Oguntuase, 2021) through collaborative projects and investments in mitigation and adaptation strategies. Such collaborations between the Global North and South are based on "long lasting concerns" about the anthropocene, leading to the emergence of the global climate action and energy security regime. Therefore, it is since the last decade that new approaches towards climate change, emission reduction and energy security through renewables have gained ground which is reflected in the emergence of global partnerships like the India-led International Solar Alliance (ISA), Coalition for Disaster Resilient Infrastructure (CDRI) (Hristova and Chankova) and the EU Green Deal. The analysis of India as a case study on promoting renewables as a climate change mitigation strategy and securing its energy requirements through self-reliance indicates its global image transformation: from being a postcolonial state relying heavily on coal-based fuels to an emerging economy inspiring the Global North countries with its ambitious climate targets to be achieved by switching over to renewables. Therefore, it would be correct to say that India's "proactive and cooperative internationalist approach in recent climate engagements" reflects its position as a responsible global partner in renewable energy investments and as a "global powerhouse" of renewables, given its high solar energy and green hydrogen potential. India's ambitious renewable energy targets towards climate resilience are wider research issues since they no longer remain confined to working with either emerging powers, or the Global South or the Global North in international collaborations (Thakker, 2018). India is a member of many significant multilateral groupings—the G20,[5] the G-77,[6] the Bay of Bengal Initiative for Multisectoral Technical and Economic Cooperation (BIMSTEC),[7] the Brazil, Russia, India, China and South Africa (BRICS) and the Brazil, South Africa, India and China (BASIC) that was instrumental in the UN Conference of Parties (CoP) held at Copenhagen and Paris where India announced the creation of the International Solar Alliance (ISA), apart from submitting its Intended Nationally Determined Contributions (INDCs) that focused on emission reduction through renewable energy (Harvey and Goldenberg, 2015). Hence, India's ambitious renewable energy targets kick-started in the Paris Agreement through the ISA and CDRI on the international front and the INDCs on the domestic front.

INDIA'S INTERNATIONAL COMMITMENTS TOWARDS NET-ZERO

India, in its initial stages of climate change negotiations, clearly articulated its position, urging the international community to acknowledge the "historical responsibility of the North" for its emissions (Dubash, 2013). Consequently, at the United Nations Framework Convention on Climate Change (UNFCCC) Rio Summit of 1992, India worked with other members of G-77 countries, urged the Global North countries to take initiatives on global climate change such that the Third World countries would have to take responsibilities voluntarily in terms of transfer of finance and technologies from the industrialized countries (Dasgupta, 2012). As a result, the summit set an "equitable agenda" and institutionalized the principles of "Common but Differentiated Responsibilities"[8] (CBDR) in Rio Summit of 1992 (Hurrell and Sengupta, 2012) (Thaker and Leiserowitz, 2014). This was India's first diplomatic victory and was further channelized in the 1997 Kyoto Protocol which confirmed the Global North's initiatives of reducing their emissions while India and other Global South nations were exempted from bearing the burden of climate change mitigation costs. The world's largest democracy has, of late, been an "emerging economic power"[9] and adhered to the principle of CBDR. The definition of CBDR was further elaborated at CoP 25[10] in Madrid in which India reiterated its claims that the developed countries were primarily responsible for the ecological damages caused to the developing and underdeveloped nations because of their rapid industrialization and that they should take a lead in making up for the damages by "fulfilling climate finance commitments" of 100 billion dollars annually by 2020. Therefore, the developed nations have "deeper pockets of responsibilities." India further argued that the procedures should be planned such that the developing countries do not bear additional burden to fulfil "pre-2020 commitments" that stands for the promises to be fulfilled by the Global North nations under the Kyoto Protocol. Later on, it was during the CoP 21 in 2015 that the loss-and-damage (L&D) fund mechanism (Warsaw International Mechanism)[11] was established and was formally agreed to by the developed nations at the CoP 27 held at Sharm-el-Sheikh after India reiterated on an L&D framework, a formalized realization of the "deeper pockets of responsibility" (Oguntuase, 2023) of the developed nations as argued earlier. The L&D fund, a historic deal reiterated by India at the UN CoP 27 in 2022, focuses on "the inclusion of 1.5°C temperature limit" (Harvey and Goldenberg 2015) for the developing nations through the cooperation of the developed world in which the developed countries, responsible for climate change, would "compensate vulnerable countries suffering from climate change" (Wyns 2022). Thus, India was seen as a "cooperative" (Oguntuase, 2021) partner and a "climate-enthusiast" that "staked" (Oguntuase, 2021) the Global South's claims of "climate reparations" which the developed nations owe to them.

However, it has been in the late twentieth century that there is a "renewed focus" on climate change through measures like afforestation, reducing reliance on conventional fuel sources and switching over to renewables. This has been accelerated during COVID-19 when global supply chains were disrupted that affected universal access to electricity, especially in the Global South.[12] It is in this backdrop that India has widened its global outreach to solve both climate and energy security multilaterally through its "grassroots' level ratification" of international climate commitments. Such an opportunity is offered by G-20 whose presidency is wielded by India between December 2022 and 2023 respectively. Of late, India has become an inspiration for both Global North and South, given its ambitious climate targets and their achievement before deadlines, thereby, offering the lowest prices of solar-powered renewables and green hydrogen at competitive prices globally superpower[13] and setting an example of sustainable development. Such achievements prove India's potential in leading global renewable energy projects which adds to employment opportunities, strengthens economies and enables it to become a global exporter of RE. It has invited leading global energy players energy players to invest in India's renewable energy projects that are mutually beneficial.

India announced that it had already achieved its Intended Nationally Determined Contributions (INDCs) that every country is required to achieve according to the Paris Agreement.[14] Its first set of INDCs, announced in 2015, aimed towards achieving its 40 percent of its cumulative electric power requirements from non-fossil fuel sources and "reducing its emissions intensity of GDP by 33 to 35 percent"[15] by 2030 as compared to 2005 levels. However, much before 2030, India achieved its targets of fulfilling 40 percent of its installed electricity from renewables in 2021 that calls for India's pursuit of advocating the "energy-economy" dichotomy through Mission LiFE (Lifestyle for Environment) and the five-fold *Panchamrit*[16] (five nectar elements) strategy, focused on renewables. Therefore, it was in 2022 again that India announced its updated INDCs to be submitted to the UNFCCC through the achievement of the *Panchamrit* goals announced at CoP-26 in Glasgow to reach net-zero by 2070. In it, it has increased its emission intensity up to 45 percent of its GDP to be reduced (10 percent more than its previous INDCs) and up to 50 percent of its cumulative electric power installed capacity from non-fossil fuels (10 percent more than what was outlined in its previous INDC clauses) by 2030. These reflect India's continuous adherence to the principle of Common but Differentiated Responsibilities with Respective Capabilities (CBDR-RC) and its commitments towards net-zero along while achieving SDGs through clean energy initiatives. Hence, the updated framework needs more technological know-how and more investments to increase manufacturing renewable energy and low emission products like electric vehicles and green hydrogen that further enhances employment opportunities for the locals at the grassroots level and international exports, at the macro level. At this stage, what India needs is finance and technology transfer from the developed countries to address its climate and renewable energy targets as reiterated by the L&D fund and such targets will drive international cooperation towards achieving renewable energy targets.[17]

It is since 2014 that the Indian economy is the world's fastest growing energy hub in terms of "renewable capacity transition" (Anand, 2023). To have achieved 40 percent of its installed electricity capacity through renewables in 2021, nine years before its deadline that indicates its ambitious and consistent global efforts towards net-zero. Its ethanol-blended petroleum initiative, that started off with 10 percent and is now working towards 20 percent blending, has sought global attention since it was scheduled to be achieved by 2030 but has been preponed between 2025 and 2026 (Anand, 2023). As India imports 85 percent of crude oil and 55 percent of natural gas, high fuel prices challenge its economic and energy security which has led to its attainment of self-reliance or *Atmanirbharta* in renewable energy transition and engaging in multilateral partnerships with both the Global North and South (Anand, 2023). Its National Green Hydrogen Mission, announced in January 2023, has captured the attention of the United States and EU, in general, and European countries like Germany and France, in particular. India's target to achieve 5 MMT green hydrogen not only requires huge investments but also manufacturing electrolysers, green steel and long-haul fuel cells that opens up a plethora of investment opportunities for the West.

Presently, India is the world's third largest producer of RE,[18] thereby, ranking fourth in both installed wind and solar power capacity.[19] India's ambitious energy transition projects, that aim for self-reliance or "*Atmanirbharta*," are inclusive, ambitious, action-oriented and significant for both the developed and developing countries that direct its geopolitical footing in the post-pandemic world as a "global superpower of green energy." PM Modi has referred to India's green energy potential as a "gold mine or oil field" (Anand, 2023) that shall bear high yield to investors that indicates how India is willing to collaborate with both domestic and international players to boost its green economy, based on solar, hydro and wind energy projects. With its highly ambitious climate targets for net-zero within 2070, India's Budget for the financial year 2023–2024 has rolled out its extensive plans to "establish India as lead player in the global green energy market" with its three pillars, namely, reducing reliance on fossil fuels, enhancing renewable energy production and focusing on a "gas-based" economy that works towards "exacerbating fuel poverty" and facilitate a just, stable and sustainable energy transition. It allocates 35,000 crore INR (4500 crore US dollars approx.) for green energy investments which is a handsome amount for an emerging Global South economy.

The India Energy Outlook 2021, released by International Energy Agency (IEA) ranks India as "the world's third-largest energy consuming country,"[20] China and the United States being at the first and second positions respectively[21] (as per BP Statistical Review of World Energy published in 2022). Achieving net-zero targets by 2050 is indeed a challenging task for 130 countries and requires multilateral engagement for its energy transition goals. Manchanda (2022) regards India's net-zero target achievement as an "uphill task" (Manchanda 2022) since its energy needs were primarily met by coal and other fossil fuels since its independence and renewables like solar and wind energy are dependent on direct sunlight availability and wind speed for a given time that impart instability to the power sector. Moreover, these are highly expensive projects that raise the prices of electricity (both in domestic and commercial sector) in the long run. Therefore, India has initiated the National Green Hydrogen Mission and the GOBARdhan

(Galvanising Organic Bio-Agro Resources-dhan) project, the former utilizing large amounts of water and the latter animal, kitchen, agricultural and other biodegradable waste to produce electricity. The GOBARdhan (Galvanising Organic Bio-Agro Resources Dhan) utilizes cattle and biodegradable waste under India's landmark rural cleanliness drive, the Swachh Bharat Mission Grameen that ensures village cleanliness, prevents air pollution due to farm stubble burning (Manchanda, 2022) and generates energy from these wastes, along with enhancing employment opportunities for the rural population. The GOBARdhan, on a larger scale, has been implemented as SATAT (Sustainable Alternative Towards Affordable Transportation) by the Indian Ministry of Petroleum and Natural Gas to motivate "entrepreneurs to set up CBG plants, produce and supply CBG to the fuel marketing companies" as domestic, industrial and automotive fuels. In the annual budget of 2023–2024, the Hon'ble Finance Minister Dr. Nirmala Sitharaman has announced the establishment of 500 new "waste-to-wealth" CBG (Compressed Biogas) plants to promote the GOBARdhan initiative in running a "circular economy"[22] where the wastes are subjected to "anaerobic digestion" (absence of oxygen) and the obtained slurry, known as Fermented Organic Manure (FOM) is used as fertilizers while the released gas (biogas) is used as "a clean cooking fuel like LPG for lighting and electricity generation" (Singh and Jawed, 2023) purposes. This is known as Compressed Biogas (CBG) and it can be purified up to 98 percent for domestic and commercial purposes. It is a renewable gas and, although possessing same components as that of CNG chemically, is capable of replacing CNG in automobiles which is produced from biomass, unlike the latter that is derived from petroleum. Consequently, CBG becomes an environmentally viable option to replace CNG for commercial, industrial and transportation purposes under the SATAT initiative (Trivedi and Srivastava, 2021). Therefore, India's concerted efforts towards biogas promotion indicate its future possibilities of replacing imported coal-based fuels and LNG that would be cost-effective, reduce agricultural waste burning, and employment generating. Hence, the initiative reflects a positive correlation amongst waste management, employment generation, energy security and organic farming. States like Haryana, Gujarat, Maharashtra and Punjab have utilized their agricultural wastes under the GOBARdhan initiative that indicates Centre-state cooperation in waste management and energy security.

Similarly, green hydrogen is something which holds a key role in India's transition to renewables and, in the long run, possesses the potential to substitute fossil fuels for transportation like shipping. Obtained through electrolysis, green hydrogen does not emit GHGs and is useful for industrial decarbonisation. The National Green Hydrogen Mission was approved on 4 January 2023[23] wherein the government aims to achieve five million metric tonnes of green hydrogen within 2030, an ambitious target for a Global South country in the emerging world order. In the global race for renewables, India has become a global market of Green Hydrogen which influences India's global leadership to attain energy self-reliance, given its "large landmass and low renewable energy costs" (Kumar, 2023). Till now, costs of electrolysers were high that upsurges green hydrogen production costs but with different waivers in government taxations, the reduced transmission and distribution costs will help the manufacturing industry in states like Kerala, Odisha, Rajasthan, Uttar Pradesh and Tamil Nadu (Kumar, 2023). This enables India to save its import costs and reduce its carbon footprints that are essential to attain net-zero. The Mission is expected to help India reduce 45 percent of its GHG emissions by 2030, as outlined in its updated INDCs, to attract FDIs in electrolyser manufacturing that ultimately reduce their manufacturing costs, help in industrial decarbonization and, consequently, boost exports for which Green Hydrogen Hubs have been developed that work on PPP framework (for example, Indian Oil and Adani Enterprises).[24] for technological innovations, research and development to create an "emerging, globally competitive green hydrogen economy." Therefore, in the near future, it is expected that green hydrogen will "enable the emergence of a domestically produced energy carrier" to reduce natural gas, petroleum and fertiliser import dependency.

INDIA'S MULTILATERAL PARTNERSHIPS— LINKING GLOBAL NORTH AND SOUTH

India's entry into the United Nations Security Council (UNSC) as a non-permanent member marked the birth of "a reformed m (Vachhatani 2020) as outlined by its "Five S" vision of "*Samman*" (Respect), "*Samvaad"* (Dialogue), "*Sahyog*" (Cooperation), "*Shaanti*" (Peace) and "*Samriddhi*" (Global Prosperity). India's notions of cooperation and global prosperity are marked by its ISA and CDRI initiatives. The ISA was born in 2015 under India's ambitious "One Sun, One World, One Grid" (OSOWOG) initiative, announced by PM Modi at CoP-21 and the-then French President François Hollande, that reiterates India's importance as a global partner in climate action and to promote renewables, in particular. As a multi-nation membership organization, it highlights the potential of the "sunshine countries" or the *Suryaputras* (the countries lying fully or partially between the Tropics) and was ratified by the United Nations in 2017 (Oguntuase, 2021). Headquartered in India, the ISA facilitates global ambitions of net-zero by promoting the use of solar energy in domestic and commercial purposes, agriculture, health, transport and electricity generation through investments' mobilization. For this, it has partnered with multilateral development banks (MDBs) and development financial institutions (DFIs), thereby, actively engaging both the public and private sectors, civil society and other international financial and technical institutions to deploy cost-effective technological know-how in the Least Developed Countries (LDCs) and the Small Island Developing States (SIDS).[25] The ISA platform promotes solar energy as a viable renewable through cooperation, technology transfer, and financial investments to achieve net-zero targets globally. It not only reduces emissions but improves energy access, increases employment opportunities for the rural livelihoods and makes energy "affordable and universally accessible by 2030" (Oguntuase, 2021). Similarly, the Coalition for Disaster Resilient Infrastructure (CDRI) was launched by PM Modi in 2019 UN Climate Action Summit that brings together both the Global North and South on a single platform. A forty-membered (Saxena, 2023) multilateral global alliance in climate action and disaster resilience, the CDRI comprises member countries, UN agencies and programmes, private sector companies, MDBs and other financial institutions that work for developing resilient infrastructure for technological innovations, recovery and reconstruction and capacity-building programmes in developing countries and Small Island Developing States (SIDS), especially in the Pacific region, that are climate-vulnerable like Singapore, Fiji and Cook Islands.[26] Hence, India's pragmatic multilateral engagement in renewable energy is reflected in its leadership in ISA and CDRI that adhere to its 5S principles and elevate its international status as an emerging global RE hub that links both the Global North and South.

Of late, India has become a global hotspot for investors in renewable energy projects like "a whiff of fresh air" (Sinha, 2020) because of its highly ambitious clean energy deployment practices (the Panchamrit goals and its submission of second set of INDCs in 2022) and the continuous support from the Indian government to provide cheap energy access at minimum costs which makes it one of the top-five clean energy producers. These have attracted multilateral and bilateral agencies and various financial institutions to invest in the nation's solar and wind energy projects, electric vehicles and energy production from wastes. Therefore, India's continuous pursuit of net-zero targets achievement requires more investments in the RE sector, that has possessed an "untapped investment potential" over the past decades. Government steps to induce Foreign Direct Investments (FDIs) have witnessed a rise during COVID-19 because of which it ranks 63rd in the World Bank's Ease of Doing Business Index 2022,[27] 37th in the Institute for Management Development's (IMD) World Competitiveness Index[28] and in 2021, the EY RE Country Attractiveness Index ranked India in the 3rd position[29] which show how its ranks have "steadily improved" as it allows 100 percent FDI, technology transfer and tax breaks in its mega RE projects, thereby, helped lowering investors' risk perception. Consequently, global investors like Abu Dhabi Investment Authority, Brookfield, CPDQ, Goldman

Sachs, JP Morgan Chase and Morgan Stanley have entered the Indian RE market that reflect how India's RE projects have become globally popular and become fertile ground for strengthening multilateralism (Sinha, 2020).

Such improvements in RE investment sector attract Western countries like the United States and European countries. In 2022, India and the United States signed the U.S.-India Strategic Clean Energy Partnership (SCEP) to "strengthen and accelerate clean, secure and just energy transition" in back-drop of COVID-19 where both the sides reiterated their commitments to boost renewables to achieve climate change mitigation goals in the age of "volatile global energy markets." These would be ensured by enhancing technological transfer, innovative solutions and industrial decarbonization, joint R&D on electric vehicles, smart grids, deployment of clean energy technologies, energy storage and Carbon Capture, Utilization and Storage (CCUS), classified under the U.S.-India Partnership to Advance Clear Energy-Research (PACE-R). This calls for greater private sector engagement to mobilize investments and technology deployment. Consequently, the Energy Storage Task Force and Low Emissions Gas Task Force have been formed that work on public-private partnership (PPP) to "support large scale integration of renewable energy" especially methane abatement technologies for emission reduction, green hydrogen and biofuels.[30] The SCEP reflects increasing bilateral energy trade of both the countries and greater investment with opportunities for capacity-building, technological research and development. It is a great opportunity for India to engage with the Global North to balance global energy markets along with achieving net-zero targets. India has solar energy, biofuels and biogas to offer to the West while the West has its technological know-how to offer to the East. The partnership helps India achieve its 50 percent of cumulative installed electricity capacity from renewables within 2030. Recently, the US Department of Energy announced that India could attain "energy independence by 2047"[31], that is, it would celebrate its platinum jubilee of political independence (from British colonialism) by minimum reliance on fuel imports and, consequently, boosts consumer savings. According to a study by the Lawrence Berkeley National Laboratory,[32] India shall achieve its net-zero targets ahead of 2070 and its achievement of complete self-reliance or *Atmanirbharta* requires an investment of 3 trillion dollars in the years to come that aims to boost green energy in the economy, that results in greater economic, energy and environmental benefits and increases its global industrial competitiveness. Having installed more than 500 gigawatts of renewable energy capacity before 2030, it is expected that India will achieve "100 percent vehicular electrification by 2035, complete industrial decarbonisation, 80 percent of clean grid by 2040 and 90 percent by 2047." The successes of the "*Panchamrit*" are reflected in its strategic partnership with the United States on climate change mitigation measures. In 2021,

India and the United States launched the Climate and Clean Energy Agenda 2030 Partnership. Its important track, namely, the Climate Action and Finance Mobilization Dialogue (CAFMD)[33] was launched by the Indian Union Minister of Environment, Forest and Climate Change Bhupendra Yadav and the US Special Presidential Envoy for Climate (SPEC) John Kerry. India was applauded by Kerry for having achieved 100 gigawatts of renewable energy much before 2030 and assured of US assistance through the ISA, climate finance and the Mission for Climate Ambitions and Agriculture Innovation for Climate (AIM4C). The U.S.-India Strategic Energy Partnership (SEP) outlined the future prospects of research in clean energy technologies and critical carbon power cycles through Carbon Capture, Utilization and Storage (CCUs). These partnership agreements focus how they are concomitant with US' FIRST (Flexible, Innovative, Resilient, Small, Transformative) Initiative and India's commitment to the Paris Agreement (Mohan, 2017).

As the world struggles to recover from the widespread economic and environmental devastation during COVID-19, the Indo-EU (European Union) cooperation on green development calls for a "renewed, global, green multilateral" (Jaspal, 2022) approach towards attaining RE targets for net-zero. The EU constitutes India's third largest trading partner and occupies the second-largest position as India's export destination.[34] "The EU-India Strategic Partnership: A Roadmap to 2025" was signed between the two in 2020 and joint statements were issued to strengthen engagement under the Clean Energy and Climate Partnership. Consequently, in December 2021, the Energy Panels of both India and the EU agreed to work towards implementing the EU-India Clean Energy and Climate Partnership involving the operationalization of the Paris Agreement till 2023. It includes cooperation in renewable energy, zero energy utilizing concrete structures, multilateral partnership through different forums that include the ISA, investments in green hydrogen, development of smart grids and climate financing. It was for these reasons that India issued Green Bonds[35] worth 300 million euros by Power Finance Corporation Ltd. (PFL) in 2017 for seven years which is the leading Non-Banking Financial Company (NBFC) in India's power sector. Such partnerships clearly indicate India's adherence to its "*Samvad- Sahyog- Samriddhi*" approach that enables it to work with a vital Global North economy, according to its *Panchamrit* principles to achieve net-zero. The India-EU cooperation on Green Hydrogen is an example that is fundamental to the hydrogen policies of both the parties, which leverages "private sector investments in innovative technologies" as discussed by the EU Global Gateway Initiative.[36] The EU Global Gateway Initiative was launched under the French presidency of EU in 2021 to bring together all the emerging markets of EU's partner countries, with India being one of the most important trade partners[37] in assisting EU's international commitments towards energy and ecological transition by promoting renewables.[38] Such cooperation is reflected through the India-EU Green Hydrogen Forum,[39] formed in 2022, that comprises the Indian Ministry of New and Renewable Energy, the Delegation of the European Union along, Hydrogen Europe and Confederation of Indian Industries (CII) which serves as a platform for both the parties to invent innovations and research in international hydrogen trade, international standards of sustainable transportation and joint projects on hydrogen. Further agreements have been concluded between the EU and India under the EU-India Clean Energy and Climate Partnership (CECP) to exchange clean energy technology and promote R&D on renewable energy solutions like offshore wind energy, solar power plants, storage and supply of RE as a sustainable climate change mitigation strategy and to achieve energy security.[40] This includes the EU joining the India-led ISA to promote solar energy which reflects North-South cooperation on RE with the former aspiring to achieve a net-zero by 2050.[41] To facilitate investments in RE, the EU Investment Bank (EU's bank) has joined another India-led initiative, the CDRI, in which it has offered to assist the Global South in its climate action strategies through the EU Green Deal and Global Gateway initiative and to build "climate change-resilient global economy" by financing RE projects. Its membership facilitates technology transfer, standard and certification development for disaster-resilient infrastructure. Concerted towards the developing world, the EUIB's climate finance is important as part of the L&D fund argued by India at UN CoP 27. Therefore, the CDRI will work with the EUIB to significantly reduce the climate vulnerabilities of the marginalized communities in the developing and underdeveloped countries towards sustainable and green development (Roy Chaudhury, 2022b). To sum up, the ISA and the CDRI are fundamental to Indo-EU CECP that meets the energy requirements of India's ever-increasing population and the EU's initiative to solve energy security challenges in a through renewables sustainably to achieve net-zero targets respectively.

Though India and the EU have cooperated on RE, India has signed several bilateral agreements with other European countries like France, Germany, Italy and the Nordic countries (Vestre and Skinnari, 2023). Since 2015, India and France have had close partnerships on climate change and renewable energy projects through the foundation of ISA by both of them. The ISA, founded by India and France, works on RE sources like biomass, solar, wind and green hydrogen. With the coordinated support of PPP, France "endeavours" to support India's transition to clean energy and achieve its net-zero targets by "fostering synergies" in energy-efficient power systems, industrial decarbonisation, green financing and boosting renewables. President Macron has also agreed to partner with India to achieve India's ambitious Panchamrit targets as set by PM Modi at CoP-26 in Glasgow. In 2022, both agreed to work on a roadmap to green hydrogen that would consequently develop "decarbonised hydrogen value chain,"[42] thereby, fostering technology transfer, research and development between Indian and French energy industries on PPP mode, that has led to a strong Indo-French strategic partnership in renewable energy, an example of India's green energy projects with a European country bilaterally.[43] With France and India having announced their net-zero targets by 2050 and 2070 respectively, it is impertinent for both to opt for complete emission-free economy through sustainable urban development, industrial decarbonization and promotion of electric vehicles. With India's Roadmap for Green Hydrogen being an inspiration for France's transition to a green economy. Hence, in 2022, the Indo-French Roadmap on the Development of Green Hydrogen[44] was adopted to foster bilateral ties on green hydrogen in order to decarbonize vehicles and industries. Manufacture of green hydrogen and electrolysers, storage, transportation, utilization and knowledge-sharing on green hydrogen are key areas of the framework which are to be implemented in the industrial, transport, energy and fertilisers' manufacturing sector that contribute to higher carbon footprints. Therefore, decarbonized hydrogen industry, environment conservation and renewable energy is a prospective area of interest for Indo-French ties towards attaining net-zero. Germany is India's most crucial partner in Europe in green energy. Recently, it has promised a sum of 10 billion euros[45] and 1 billion euros later (Laskar, 2022) in India's climate action through its energy transition projects, concluded the Indo-German Renewable Energy Partnership that focused on boosting solar energy and renewables towards a just transition.[46] Before that, it had concluded the Indo-German Partnership for Green and Sustainable Development wherein both "agreed to develop an Indo German Green Hydrogen Roadmap"[47] that promised technology transfer, R&D in green hydrogen manufacture that limits the temperature rise up to 2°C above pre-industrial levels through emission reduction. Furthermore, the Indo-German Renewable Energy Partnership and the Global Initiative for Disaster Risk Management[48] (a similar initiative like the CDRI) focus on boosting renewables like solar and wind energy projects for a "Just Energy Transition Partnership" (JETP) through finances and PPP in technological innovations on their way to sustainable development by finding India's more than 20 development projects on RE (Laskar, 2022). Negotiated by Germany, the JETP supports India's ties with the G-7[49] countries and the United States in renewable energy and green hydrogen production (Laskar, 2022). This makes India "the single largest recipient of German cooperation" in phasing out coal fuels and has helped India cut down over "100 million tonnes of CO2 in recent years" (Laskar, 2022). Meanwhile, Denmark seeks to reduce 70 percent of its GHG emissions by 2030 and has, therefore, signed the Indo-Danish Green Strategic Partnership in 2020 to strengthen cooperation with India for green growth, emission reduction, job creation for the successful implementation of the Paris Agreement commitments.[50] Another RE Partnership with a European country, the Indo-Italy Strategic RE Partnership, signed in 2021, works on areas like distribution networks of RE, technology transfer in manufacturing green hydrogen and new solar solutions, alternative technologies on mobility like electric vehicles, green hydrogen fuel cell mobility, biofuels and waste fuels (as initiated under the GOBARdhan scheme), innovative waste management strategies and sustainable smart cities on PPP mode with the Indian public sector units like Indian Oil and Italian energy companies like Piaggio, Snam and Enel Green Power (Roy Chaudhary 2022a).[51] Therefore, with Italy, Denmark and other Nordic countries having concluded various bilateral green energy agreements with India, it can be safely concluded that France and Germany are India's most important European partners on green cooperation and multilateralism (given their continuous support to the India-led ISA and CDRI), sharing "non-partisan commitments towards growing and deepening bilateral relationship" (Laskar, 2022) to rejuvenate synergies between the two parties for building self-reliance and sustainability, especially in the post-pandemic world where climate and energy crisis are knocking at the doors.

Africa, possessing an "arguably limitless" (Oguntuase, 2021) solar energy capacity given its geostrategic location, needs to utilize it to the fullest. Lying in the Tropical region, the African continent possesses huge potential of solar energy with some of its countries being members of G-20 while all its countries constituting the G-77 (Oguntuase, 2021). Meanwhile, under the CBDR, the G77+ China coalition has helped Africa to mitigate climate change risks, as the L&D, asserted by India, enunciates. Therefore, Africa offers ample opportunities for South-South cooperation (Lessole, 2014) in economic and technical spheres under the United Nations Framework Convention on Climate Change. Having realized Africa's potential, India has invested in its solar projects under the aegis of the ISA. Spearheaded by India, African countries like Ghana, Madagascar, and South Sudan are members of the ISA. The ISA and the African Development Bank have pledged to develop solar power systems in the Sahel region of Africa (Kaul, 2019) to provide electricity. Other countries like the Republic of Congo, Mali, Burkina Faso, Nigeria, Uganda, Tanzania, Nigeria, Senegal and Ghana (Oguntuase, 2021) have shown keen interest in partnering with the ISA to share solar energy deployment practices. Such solar projects include rooftop solar panels, solar mini-grids, solar PV power plants, rural electrification, street lighting, solar cooling systems, solar-powered irrigation systems and urban infrastructure like schools, colleges, hospitals and public establishments. It has contributed significantly to poverty alleviation, job creation leading to increased incomes, access to quality education and healthcare, social security and towards ecological sustainability by reduction in deforestation, land degradation and increase in land reclamation. These reflect India's dynamism, under the ISA, in Africa's solar power plants and green technology transfer, followed by bilateral trade and investment, scientific and technological cooperation. As outlined by PM Modi in the ten Kampala principles of India-Africa engagement in 2018 to prioritise Africa's development and foster South-South cooperation (Viswanathan and Mishra, 2019), India's climate change mitigation strategies and renewable energy projects in Africa rejuvenate its historical ties with the latter, both belonging to the Global South. This indicates that as developing countries, both need finance from the developed world to assist them in their mitigation and adaptation strategies (Viswanathan and Mishra, 2019) in their net-zero targets, leading to a South-South cooperation.

Recently, the ISA, the Grid Controller of India and the West African Power Pool (WAPP) held a meet of African countries to exchange ideas on solar energy implementation. This included 60 members from 13 WAPP countries to plan solar energy power generation framework capacity, knowledge-sharing and capacity-building to improve people's living standards and ensure clean energy access at minimum costs in developing countries. Prioritizing solar energy generation, it aimed to set global standards of "solar ecosystems in all ISA member countries"[52] to fight against climate change and mitigate energy crisis. The ISA, along with the United Nations Development Programme (UNDP), has launched various solar powered projects to improve irrigation facilities for agriculture like solar water pumping systems and capacity-building for the same in African countries like Benin, Mali, Nigeria, Senegal, Republic of Sudan and South Sudan to provide cost-efficient energy, reduce carbon emissions and add to farmers' incomes (Gupta, 2023). It now looks towards expanding its membership with Africa's inclusion for a self-reliant sustainable disaster-resilient infrastructure. Its IRIS (Infrastructure for Resilient Island States) initiative, launched at CoP 26, works towards resilient infrastructure development in Australia, Fiji, Jamaica and Mauritius that are climate-vulnerable which restricts their economic development and impoverishes it (Saxena, 2023). Recently, the EU joined the CDRI in the backdrop of COVID-19 when the world needed a robust climate framework, amidst the renewed need for investments in climate change mitigation strategies, renewable energy, green infrastructure and resource efficiency needed for a "circular economy" that are some thrust areas of cooperation under the EU-India Strategic Partnership and are outlined by the Paris Agreement towards GHG emissions reduction for enhanced climate change adaptation. With hydrogen production being one of the pillars of global energy transition towards net-zero, its utilization in CDRI "at the current nascent stage can yield exponential benefits" in the age of climate change disasters, although its manufacture accounts for huge availability of water resources amidst the changing climatic conditions which could be effectively utilized under the EU Green Hydrogen Mission. According to the 2022 Intergovernmental Panel on Climate Change

(IPCC) Report on Climate Change Mitigation, hydrogen is a carbon-neutral source of electricity in heating, heavy industry and transport.[53] Therefore, as the leading country, India seeks to ensure investments in renewable energy to reduce GHG emissions, phasing out coal and other disaster mitigation strategies (Roche, 2021) to achieve its net-zero goals.

These are also furthered in its other multilateral disaster management mechanisms, such as the Agreement on South Asia Rapid Response to Natural Disasters (2011), the BIMSTEC Centre for Weather and Climate (2014), the South Asian Cooperative for Environment Protection (2018) and the establishment of the BIMSTEC Technology Transfer Facility (BIMSTEC-TTF) in 2022 have strengthened international cooperation in climate change mitigation measures and renewable energy for ports and shipping industry. The initiatives reflect India's "emerging focus" on multilateral engagement to tackle environmental emergencies that the neighbourhood faces and affects it too. Such measures "strengthen joint disaster response mechanisms" through prevention and mitigation plans so as to rapidly respond to the climate disasters without further damages, promoting renewable energy for all projects and attracting climate finance through the L&D fund as the region is geographically one of the worst-suffering locations of the Global South. India's Act East Policy, that strives to strengthen ties with the ASEAN (Association of Southeast Asian Nations) countries have been clubbed as its "extended neighbourhood." These nations are integral to the Global South for their geostrategic location near the Pacific Ocean. Lately, India organised a High-Level Conference on Renewable Energy with the ASEAN countries focusing on "Experience and Innovations for Integrated Renewables Market" where Indo-ASEAN cooperation on facilitating renewable energy transitions to mitigate climate change was highlighted. India applauded ASEAN's decision to join the ISA through which knowledge-sharing, technology transfer and renewable energy infrastructure development would gear up grid integration under the OSOWOG initiative, based on Indo-ASEAN historical and cultural ties (PIB, Feb. 2022). Meanwhile, the Quad leaders, with Japan (one of the ASEAN countries) as a vital partner, launched the "Quad Climate Change Adaptation and Mitigation Package" (Q-CHAMP) in 2022 under which they also agreed to "implement the Paris Agreement and deliver on the outcomes of CoP 26 to strengthen climate cooperation through climate finance by both private and public sectors in the Indo-Pacific region that facilitate research, development and deployment of innovative clean technologies. The Q-CHAMP would work on green shipping initiatives, similar to the BIMSTEC-TTF, that aimed towards a "shared green corridor framework" to strengthen renewable energy projects like clean hydrogen that reduces emissions (White House, 2022). Both BIMSTEC-TTF and Q-CHAMP are working on the regions where water is a readily available resource that is needed to produce clean hydrogen. The

Quad nations are steadfast in their efforts towards the achievement of SDGs, renewable and clean energy innovations and emissions targets of 2030. This is reflected through their focus on "improving critical climate information-sharing and disaster-resilient technical infrastructure" in the Indo-Pacific through the CDRI for technical assistance in Small Island Developing States (SIDS) and a Task Force in Climate and Information Services are certainly good initiatives (Kumar, 2022).

Furthermore, wind and hydel power projects are also best suited for the region. Meanwhile, both of them shall also work together disaster management strategies since the regions are worst affected by tsunamis, earthquakes and floods due to rising sea levels. CCUs and climate-smart agriculture are on the forefront in which the Sydney Energy Forum plays an important role. Therefore, a clear connection is revealed that India is a strategically important partner for BIMSTEC, Quad and the ASEAN because of its geopolitical location, its workforce, its potential as a "global manufacturing hub for green hydrogen" and as a fertile ground for renewable energy investments that could perhaps counter Chinese hegemony in renewables, especially solar grids and e-vehicles, in the Indo-Pacific to maintain a balance of power. Surprisingly, First World nations like the United States, Germany and Japan, already possessing electrolyser technologies, are willing to invest, import and build global supply chains of green hydrogen from India, apart from trading green hydrogen in the region. Collaboration with European countries like Denmark, Japan, Norway and Sweden, possessing good electrolyser technologies, shall reduce electrolyser costs which would, in turn, reduce the manufacturing costs of green hydrogen so that they are cost-effective as compared to the conventional energy sources. The Quad further promises to enhance clean hydrogen and ammonia markers in the region to mitigate emissions and to "participate in high-integrity carbon markets" (Kala, 2022). India's adoption of a five-S approach to multilateralism in climate diplomacy involves respecting principles like *Samman* (Respect), *Samvad* (Dialogue), *Sahyog* (Cooperation), *Shanti* (Peace) and *Samriddhi* (Prosperity) which is clearly reflected in its engagement with the BIMSTEC countries to strengthen its maritime security and rejuvenate the region's potential for blue economy in the Indian Ocean Region (IOR) through technical and technological assistance to promote climate resilient ports and renewable energy (Joshi, 2021). Focused on solar energy, offshore wind energy, biofuels, energy efficiency in buildings and green hydrogen, both of them reiterated in 2022 to "ensure the transition towards a green future" through India-led global initiatives like the ISA and the CDRI wherein more EU member states have joined the ISA.

The global initiative for Life for Environment (LiFE) movement was launched on 5 June 2022 to "encourage an eco-friendly lifestyle" and make "pro-planet people" (Triple-P).[54] PM Modi's announcement of the LiFE movement at CoP 27 has several connotations. It outlines India's stand on a cooperative framework to secure a sustainable lifestyle by boosting agro-forestry, agro farming, using electric vehicles, banking on non-fossil fuels to reduce emissions and many more. This applies to its foreign as well as domestic politics. For this, it shall work upon a four-year program to mitigate climate, energy and food security challenges that includes ethanol-blended fuels, e-vehicles, utilizing solar and wind energy, encouraging investments in green hydrogen as an "alternate energy source" through the ISA and CDRI, adhering to its guiding principles of equity, humanity, climate justice (Mohapatra, 2022) and "*Vasudhaiva Kutumbakam*" that refers to the entire world as a family. Herein, it is seen that India looks forward to measures that are economically suitable for a forest-rich nation and ensure sustainable future for not only itself but also for its allies, that is "*Sahyog*" and "*Samriddhi*" (cooperation and prosperity of all). Furthermore, India's lead in the ISA highlights its initiatives of renewable energy, climate finance, clean technology and sustainable manufacturing as a Quad nation too.

It is generally seen that while multilateral agreements on climate change, GHG emissions reduction and renewable energy are concluded on diplomatic level, grassroots level like cities and towns are often "side-lined" (Srivastava, 2023), especially in the Global South nations. According to UN reports, it is expected that by 2050, nearly 75 percent of the global population would be living in urban areas, which makes the "interplay between the local and global" (Srivastava, 2023) more "complex," with enhanced scope of local leadership globally. Realizing the role of cities in GHG emissions and climate change in the developing world, the Urban 20 (U20) engagement group was convened by India (who is the G-20 president between December 2022 and 2023) in February 2023 to provide and share innovative local solutions to tackle global challenges according to the international commitments of Paris Agreement and the Sustainable Development Goals (SDGs). These include city networks formations, exchange of best practices and technical know-hows. Initially formed in 2017, this meet was attended by city and town administrators of G-20 nations (including both developed and developing countries) to discuss "concrete measures and policies" to tackle climate change and promote the manufacture and use of renewable energy. In 2016, the EU and India have collaborated under the Clean Energy and Climate Partnership (CECP) to strengthen cooperation in clean energy for resolution of challenges like climate and energy crisis. This was followed by India joining the Global Covenant of Mayors-South Asian (GCoM-SA)[55] for Climate and Energy in 2018, an EU-funded international Urban Cooperation Programme, under which all the South Asian countries reiterated their commitments towards GHG emissions reduction in its cities, thereby, ensuring better quality of lives for their citizens and including several Indian cities like Ahmedabad, Jamnagar and Vadodara and their mayors[56] as members of this coalition. The GCoM is "world's largest city climate leadership alliance" (Srivastava, 2023) involving 140 countries and more than 12,000 signatories to tackle climate change. In line with the GCoM and the C40 Cities Climate Leadership Group, the U20 engagement stresses solving climate and energy challenges in a post-pandemic world, along with working towards the achievement of UN SDG goals within 2030 with a bottom-up approach to decision-making. India plays a key role in "re-shaping global governance and influencing decision-making" (Srivastava, 2023) between national and local governments to voice the requirements of different cities, especially in the Global South and "enhance urban leadership." In 2023, India as G-20 president, has outlined its priorities to encourage "environmentally responsive behaviour" (Srivastava, 2023), utilizing climate finance and lending recognition to best practices of grassroots identity, thereby, transforming "intention into action" by "showcasing" (Srivastava, 2023) its own cities which are willing to partner with various countries. These initiatives reflect how cities contribute to global action on climate change and India's role in a multilateral organization like G-20 to enable technology transfer and cooperation between the cities of both Global North and South for a better future.

In 2023, the G-7 countries reiterated their commitments towards "coal phase-out" through "carbon-free electricity production by 2035" when the Indian Ministry of Environment Forest and Climate Change Shri Bhupender Yadav, who was invited as EU's most important RE partner, argued for a "phase-down" instead of "phase-out" since the developing world is "significantly dependent on coal for electricity." Hence, all agreed to mobilize more investments in the RE sector to limit the temperature up to 1.5^{0}C of pre-industrial levels by the end of twenty-first century and to assist the developing world with technological know-how, finances and investments for GHG emissions reduction to combat climate change, according to the "equity" principle of climate justice. Consequently, G-7 emphasizes on renewables to achieve energy security and accelerate clean energy transition to "achieve a "predominantly decarbonized power sector" by 2035 (Koshy, 2023). India, having attained this year's G-20 presidency, has convened the "Voice of Global South Summit 2023," a forum to discuss political, environmental, cultural and socio-economic issues of the developing world, stressed on sustainable development and energy security that calls for partnerships amongst the developing world to access green energy. Its G-20 presidency voices the Global South's concerns of "energy security, energy justice and sustainable energy transition"[57] that are to be solved by promoting renewables internationally, knowledge-sharing and investment mobilization. This includes usage of indoor solar cooking system, biofuels, ethanol-blended fuels, biogas, electric vehicles and green hydrogen in which India has achieved landmark successes.

The ISA, CDRI and the Global Biofuel Alliance under G-20 are some of the platforms through which India aims to strengthen renewable energy cooperation. As discussed earlier, India has launched its E-20 initiative of blended petroleum, the technology of which it seeks to share with other countries too. To enhance cooperation and engagement on biofuels, it will initially work together with Brazil and the United States (the leading global biofuel producers) in the form of a Global Biofuels Alliance, whose membership is expected to expand in the future. It intensifies biofuel usage in the transportation sector that strengthens energy markets, making nations self-reliant and saving import costs through the establishment of bioenergy and bio-economy.[58] The initiative outlines India's priorities as a G-20 president to enhance cooperation between the Global North and South in many spheres, biofuels being one of them. Similarly, India has recently concluded an agreement with the United Arab Emirates (UAE) to explore low-carbon technologies to limit the rise in global temperature up to 1.5ºC above pre-industrial levels.[59] Thus, both of them along with France have collaborated on solar energy to mitigate climate change and promote biodiversity conservation (Bhattacherjee, 2023). Meanwhile, India is a key partner in Saudi Arabia's "Vision 2030" for its sustainability and catalyses its aspirations to "turn Middle East into 'Europe of the Future'" (Rodrigues, 2023). Lying in the Tropical region, it possesses "impressive natural potential for solar and wind power" that is furthered by investments in technological advancements to diversify its energy resources. As a member of the Gulf Cooperation Council (GCC) and the ISA, it seeks to invest in India's Green Hydrogen Mission since the GCC countries possess affordable land and water resources to help in manufacturing green hydrogen and solar power projects (Rodrigues, 2023). Such trilateral and multilateral collaborations indicate gap-bridging between the Global North and South with India's role being that of a mediator.

CONCLUSIONS

The ISA demonstrates how India has become a "Middle Power" between the Global North and South with the United States driving capital investments through climate finance in the developing countries (Kumar, 2022). As Ms. Laura Neack rightly defines, "middle power diplomacy" refers to a nation's "tendency to pursue multilateral solutions to international problems, the tendency to embrace compromise positions in international disputes, and the tendency to embrace notions of 'good international citizenship' to guide . . . diplomacy. Middle powers are states who commit their relative affluence, managerial skills, and international prestige to the preservation of the international order and peace through coalition-building, by serving as mediators and 'go-betweens,' and through international conflict management and resolution activities, such as UN peacekeeping" (Neack, 2000). She defines it as an "imperative" that was popularised during the Cold War period. India emerged as a world leader in solar (fifth rank) and wind (fourth) in 2019, as per reports (REN21, 2020) and its leadership in ISA, G-20 and BIMSTEC brings together the developed, developing and the least developed countries together to boost low-carbon technologies since emissions and climate change are no more India's domestic challenges. It has strengthened its position by partnering with the African Union countries and with organizations like BIMSTEC, ASEAN, EU and the Quad that comprise of both Global North and South. India has always reiterated that it stands for a "cooperative" (Sahyog) approach to "peaceful" (Shanti) sustainable development for all through "dialogue" (Samvad). Therefore, it requires the concerted efforts of all the nations in the form of climate finance, technological transfer, capacity-building and utilization of natural energy resources with sustainable farming practices. This is clearly reflected in its pledge to create a corpus fund of 26 million dollars with credit lines worth 1.39 billion dollars to undertake renewable energy projects in the ISA member countries (Shaji and Susarla, 2021). India's increasing trends towards engaging in bilateral and multilateral partnerships to mitigate climate crisis and energy security challenges with its immediate and extended neighbourhood and with the Western major powers help it in earning the name of a "middle power" that solicits "inclusive" cooperation of both Global North and South countries to bolster an emission-free, green economic world order. Hence, it can be rightly said that in the coming decades, India is to "emerge as a major global climate leader" (Joshi, 2021) by guiding both developed and developing countries to generate employment, reduce GHG emissions, meet climate and green energy targets and mitigate climate disasters. Complete "coal-phaseout" remains a challenge for developing countries like India at this juncture, however.

With hydrogen production being one of the pillars of global energy transition towards net-zero, its utilization in CDRI "at the current nascent stage can yield exponential benefits" in the age of climate change disasters, although its manufacture accounts for huge availability of water resources amidst the changing climatic conditions. According to the 2022 IPCC Report on Climate Change Mitigation, hydrogen is a carbon-neutral source of electricity in heating, heavy industry and transport.[60] Climate risks posed by GHG emissions have resulted in the emergence of global partnership organizations like ISA and CDRI that focus on developing RE which reflects India's "progressive and cooperative climate engagement" (Oguntuase, 2023) coupled with promoting RE that strengthens its ties globally. Through its partnerships in Africa, India presents a positive correlation between climate change and sustainable development through the ISA's increasing contribution towards

Africa's poverty alleviation and climate change mitigation strategies, hence continuing its journey in voicing the Global South's demands. Therefore, it can be safely concluded that India has strengthened its bilateral and multilateral partnerships with both Global North and South with RE being the "glue" (Rodrigues, 2023).

NOTES

1. "GDP Is Growing Fastest in These Countries—What It Means," https://www.weforum.org/agenda/2022/01/oecd-gdp-growth-bounceback/ (last accessed 12/29/2023); "Real GDP Growth in 2021–22 stands at 8.7 per cent, 1.5 Per Cent Higher Than the Real GDP of 2019–20," https://pib.gov.in/PressReleaseIframePage.aspx?PRID=1847122 (last accessed 12/29/2023). Shri Pankaj Chaudhary was appointed as the Minister of the State in the Ministry of Finance in 2021.

2. Ibid.

3. "India Energy Outlook 2021," https://www.iea.org/reports/india-energy-outlook-2021 (last accessed 12/29/2023).

4. Intergovernmental Panel on Climate Change, Climate Change 2022: Impacts, Adaptation, and Vulnerability, IPCC, 2022.

5. Formed in 1999, the Group of 20 (or G-20) is an organisation of world's largest economies from both Global North and South including 19 countries and the European Union. It addresses global issues like climate change, energy crisis, financial stability and sustainable development.

6. Headquartered in the United Nations Headquarters of Geneva since 1964, the Group of 77 presently comprises 134 Global South economies to promote their economic interests. It was initially founded by 77 countries to address apartheid and global disarmament.

7. The BIMSTEC, formed in 1997, comprises Bangladesh, India, Myanmar, Sri Lanka, Thailand, Nepal and Bhutan, that is, the countries around the Bay of Bengal region. It is headquartered in Dhaka, Bangladesh, and meets annually to discuss about trade and investment, counter-terrorism, climate change, energy, tourism, agriculture, public health, poverty alleviation, cultural exchanges, transport and communication.

8. The CBDR principle, the first international legal instrument advocated in UNFCCC Rio Summit of 1992 to address climate change, states that all nations are obliged to address environmental destruction and climate change, however, the obligations are not equal. The developed world, being the most industrialized, responsible for climate change which allows less opportunities for the developing countries to industrialize. Hence, the developed world should pay compensation for the damages incurred and reduce their industrial emissions, hence, allowing the Global South for industrialization.

9. The world's economic hegemony seems to shift "from west to the east" and it is expected that by 2030, the top three economies in the world shall be China, India, and Japan, all being Asian. According to reports by Morgan Stanley, India is currently the

world's sixth largest economy and is the fastest growing economy in the world. It is predicted to take over Japan in Asia with its second largest projected GDP exceeding 8.4 trillion dollars and might also exceed that of Germany and UK in the world by 2030. Consequently, its per capita income is expected to cross 15 thousand dollars by 2047 with its GDP exceeding 26 trillion dollars. It shall become a global manufacturing and technology hub to diversify its supply chains, owing to its investment-attracting policies and reforms, renewable energy transitions, digitalized infrastructure, sustainable transition procedures and global competitiveness. Many Multinational Corporations (MNCs) look towards India as "an investment destination" with its "entrepreneurial, English-speaking and digitally literate" working-age population exceeding 900 million. Therefore, in the post-pandemic world, the World Bank, IMF and the Global Consulting Firms claim that India "is on the verge of becoming an important world economic power in the near future" (Deshpande, 2023) because of which it can be called an "emerging economic power."

10. "Outcome of COP25 Balanced, with the Exception of Climate Finance Issues: Shri Prakash Javadekar," *PIB,* 2019.

11. Report of the Conference of the Parties on its twenty-first session, held in Paris from 30 November to 13 December 2015, *UNFCCC,* 2016.

12. "Report: COVID-19 Slows Progress Towards Universal Energy Access," *WHO,* 2022.

13. "India Fastest in Renewal Energy Capacity Addition Among Major Economies," *ET Energyworld*, 2023.

14. "G20 Energy and Climate Joint Ministerial Meeting," *PIB,* 2021.

15. "Cabinet Approves India's Updated Nationally Determined Contribution to Be Communicated to the United Nations Framework Convention on Climate Change," *PIB,* 2022.

16. The Panchamrit principles (five nectar elements of India's renewable energy goals towards net-zero) were announced by PM Modi at CoP-26 in 2021. A mixture of five natural foods like milk, ghee, honey, curd and jaggery, the Panchamrit is traditionally offered to Gods in Hinduism and Jainism and as a medicine in Ayurveda. Therefore, the Panchamrit principles reflect India's pious ambitions towards net-zero. These include attaining net-zero by 2070, obtaining 500 gigawatts of renewable energy, meeting 50 percent of its energy requirements from non-fossil fuels, reduce its projected carbon emissions by one million tonnes and reduce the carbon intensity by 45 percent within 2030.

17. "Cabinet Approves India's Updated Nationally Determined Contribution to Be Communicated to the United Nations Framework Convention on Climate Change," *PIB*, 2022

18. "Renewable Energy in India," *PIB*, 2022.

19. "Year-End Review 2022–Ministry of New and Renewable Energy," *PIB*, 2022.

20. "India Has Been Ranked Third Largest Primary Energy Consumer in the World," *PIB Ministry of Petroleum and Natural Gas, Govt. of India*, 2022.

21. India Country Analysis Brief, *U.S. Energy Information Administration.*

22. The definition of "Circular economy," as stated by the NITI (National Institution of Transforming India) Aayog of India, aims at full utilization of resources

through recycling, gains competitiveness and minimizes resource dependency and waste generation. To facilitate these, the Government's formulate policies that promote metals recycling, plastic and e-waste management leading to self-reliance. Along with the EU, it seeks "resource efficiency in metals and e-waste." Hence, the GOBARdhan initiative is one of its effective waste management strategy. These serve many purposes: employment opportunities leading to increase in per capita income, cost-effective, ecologically sound and improving India's external ties through partnerships and collaborations in investments and technology transfer. See NITI Aayog, 2021.

23. "Cabinet Approves National Green Hydrogen Mission," *PIB, 2023.*

24. "Explained: What Is Green Hydrogen and What Is India's National Mission to Cut Emissions?," *ToI*, 2023.

25. "International Solar Alliance and West African Power Pool hosts 13 African Countries in New Delhi to Share Best Practices in Solar Deployment," *PIB*, 2023.

26. "List of SIDS," *UN Office of the High Representative for the Least Developed Countries, Landlocked Developing Countries and Small Island Developing States.*

27. "Ease of Doing Business: Govt Working on to Reduce Compliance Issues, Says Official," *Economic Times*, 2022.

28. "India Jumps 6 Places to 37th Rank on IMD's World Competitiveness Index," *The Hindu*, 2022.

29. "India Retains 3rd Position in Renewable Energy Investment Attractiveness Index," *Livemint*, 2021.

30. "U.S.-India Strategic Clean Energy Partnership Ministerial Joint Statement," *PIB*, 2022.

31. "India Can Achieve Energy Independence by 2047: US Study," *India Today*, 2023.

32. Ibid.

33. "India and US Launch the Climate Action and Finance Mobilization Dialogue (CAFMD)." *PIB, MoEFCC, 2021.*

34. "India, EU to Step Up Cooperation on Clean Energy and Climate Action," *The Print*, 2022.

35. "PFC Issues India's First-Ever Euro Green Bond," *PIB*, *Ministry of Power,* 2021.

36. "India, EU to Step up Cooperation on Clean Energy and Climate Action," *The Print,* 2022.

37. "EU-India Free Trade Agreement, Investment Protection Agreement and Geographical Indications Agreement," *European Commission.*

38. "What Is the European Global Gateway Strategy?," *French Foreign Policy, Ministry for Europe and Foreign Affairs.*

39. "EU-India Hold First Green Hydrogen Forum in New Delhi," *EEAS*, 2022.

40. "Speech by President von der Leyen at the International Solar Alliance," *European Commission*, 2022.

41. "France and India Adopt Joint Roadmap on Green Hydrogen," *French Embassy in New Delhi, India*, 2022.

42. "France and India Adopt Joint Roadmap on Green Hydrogen," *French Embassy in New Delhi, India*, 2022.

43. "French Clean Energy Days India 2023 by Business France: Fostering Indo—French Business Synergies in the Clean Energy Sector," *Business France, CISION PR Newswire*, 2023.

44. "France and India Adopt Joint Roadmap on Green Hydrogen," *French Embassy in New Delhi, India*, 2022.

45. "Germany Commits 10 Billion Euros to India for Climate Action," *Outlook India*, 2022.

46. Ibid.

47. Ibid.

48. Ibid.

49. G-7 is an intergovernmental political forum of world's largest "IMF-advanced" economies like the United States, Canada, UK, France, Germany, Italy and Japan, with the EU being a "non-enumerated" member. It addresses issues like human rights, climate change, energy security, international security and cooperation with the developing world.

50. "Green Strategic Partnership," *Official Website of the Ministry of Foreign Affairs of Denmark*, 2021.

51. "Greenko, Italy's Snam to Collaborate on Green Hydrogen," *The Hindu*, 2020.

52. "International Solar Alliance and West African Power Pool Hosts 13 African Countries in New Delhi to Share Best Practices in Solar Deployment," *PIB*, 2023.

53. "Benefits from Disaster Resilient Infra at Nascent Stage of Hydrogen Economy," *ET Energyworld*.

54. "India's Resolve for Climate Commitments Evident from Performance: PM to G7," *PTI, Business Standard*, 2022.

55. "EU and India Have a Partnership on Clean Energy and Closely Cooperate to Support their Climate and Energy Objectives," *European Commission*.

56. "Global Covenant of Mayors Launches Regional Chapter in South Asia," *International Urban Cooperation, European Union- South Asia Chapter*.

57. "Voice of Global South Summit," *PIB*, 2023.

58. "India, US, Brazil to Work Together for Global Biofuels Alliance," *Energyworld*, 2023.

59. "UAE Willing to Contribute to India's 500-GW Clean Energy Goal: COP28 President-designate Al Jaber," *Economic Times*, 2023.

60. "Benefits from Disaster Resilient Infra at Nascent Stage of Hydrogen Economy," *ET Energyworld*.

BIBLIOGRAPHY

Anand, Saurav. "India's Potential in Green Energy no Less than a Goldmine, Invest Here: PM Modi," c*Mint*, 23 February 2023. https://www.google.com/amp/s/www.livemint.com/news/india/indias-potential-in-green-energy-no-less-than-a-goldmine-invest-here-pm-modi/amp-11677143366844.html

"ASEAN-India High Level Conference on Renewable Energy Commences," *PIB, Ministry of New and Renewable Energy, Govt. of India*, 7 February 2022. https://www.pib.gov.in/PressReleasePage.aspx?PRID=1796246

"Benefits from Disaster Resilient Infra at Nascent Stage of Hydrogen Economy," *ET Energyworld*, 7 May 2022. https://www.google.com/amp/s/energy.economictimes.indiatimes.com/amp/news/renewable/benefits-from-disaster-resilient-infra-at-nascent-stage-of-hydrogen-economy/91388929

Bhattacherjee, Kallol. "India Joins UAE, France for Trilateral on Clean-energy, Counter-Pandemic Steps," *The Hindu,* 5 February 2023. https://www.google.com/amp/s/www.thehindu.com/news/national/india-joins-uae-france-for-trilateral-on-clean-energy-and-counter-pandemic-measures/article66471623.ece/amp/

"Cabinet Approves India's Updated Nationally Determined Contribution to Be Communicated to the United Nations Framework Convention on Climate Change," *PIB, Cabinet, Govt. of India,* 3 August 2022. https://pib.gov.in/PressReleaseIframePage.aspx?PRID=1847812

"Cabinet Approves National Green Hydrogen Mission," PIB, MNRE, 4 January 2023. https://pib.gov.in/PressReleasePage.aspx?PRID=1888547

"CoP26: Modi Offers 'Panchamrita' Concoction for Climate Conundrum at Glasgow," *Down to Earth*, 2 November 2021. https://www.google.com/amp/s/www.downtoearth.org.in/news/climate-change/amp/cop26-modi-offers-panchamrita-concoction-for-climate-conundrum-at-glasgow-80001

Country Analysis Brief: India. *U.S. Energy Information Administration.* https://www.eia.gov/international/analysis/country/IND

Dasgupta, C. "Present at the Creation: The Making of the UN Framework Convention on Climate Change," in N. Dubash, ed., *Handbook of Climate Change and India: Development, Politics and Governance.* New Delhi: Routledge, 2012.

Demo, Chencho. "How India's Renewable Energy Ambitions Have Wedded Bhutan to Hydropower," *Scroll.in*, 22 March 2023. https://www.google.com/amp/s/amp.scroll.in/article/1045362/indias-renewable-energy-ambitions-has-ended-up-wedding-bhutan-to-hydropower-despite-rising-costs

Dubash, N.K. "The Politics of Climate Change in India: Narratives of Equity and Co-Benefits," *Wiley Interdisciplinary Reviews: Climate Change*, 4(3), 2013, pp. 191–201.

"Ease of Doing Business: Govt. Working on to Reduce Compliance Issues, Says Official," *Economic Times*, 6 July 2022. https://www.google.com/amp/s/m.economictimes.com/news/economy/policy/ease-of-doing-business-govt-working-on-to-reduce-compliance-issues-says-official/amp_articleshow/92705974.cms

"EU-India: New Trade and Technology Council to Lead on Digital Transformation, Green Technologies and Trade," *Press Release, European Commission*, 2023. https://ec.europa.eu/commission/presscorner/detail/en/IP_23_596

"EU-India Free Trade Agreement, Investment Protection Agreement and Geographical Indications Agreement," *European Commission.* https://policy.trade.ec.europa.eu/eu-trade-relationships-country-and-region/countries-and-regions/india/eu-india-agreement_en

"EU-India Hold First Green Hydrogen Forum in New Delhi," EEAS, An official website of EU, 8 September 2022. https://www.eeas.europa.eu/delegations/india/eu-india-hold-first-green-hydrogen-forum-new-delhi_en
"EU and India Have a Partnership on Clean Energy and Closely Cooperate to Support Their Climate and Energy Objectives," *European Commission.* https://energy.ec.europa.eu/topics/international-cooperation/key-partner-countries-and-regions/india_en
"Explained: What Is Green Hydrogen and What Is India's National Mission to Cut Emissions?," *The Times of India,* 5 January 2023. https://m.timesofindia.com/business/india-business/explained-what-is-green-hydrogen-and-indias-national-mission-to-cut-emissions-/amp_articleshow/96755832.cms
"France and India Adopt Joint Roadmap on Green Hydrogen," *French Embassy in New Delhi, India*, 18 October 2022. https://in.ambafrance.org/France-and-India-adopt-Joint-Roadmap-on-Green-Hydrogen
"French Clean Energy Days India 2023 by Business France: Fostering Indo-French Business Synergies in the Clean Energy Sector," *Business France, CISION PR Newswire*, 2023. https://www.prnewswire.com/in/news-releases/french-clean-energy-days-india-2023-by-business-france-fostering-indo--french-business-synergies-in-the-clean-energy-sector-301746050.html
"G20 Energy and Climate Joint Ministerial Meeting," *PIB, Ministry of Power*, 23 July 2021. https://pib.gov.in/PressReleasePage.aspx?PRID=1738317
"Germany Commits 10 Billion Euros to India for Climate Action," *Outlook India*, 3 May 2022. https://www.google.com/amp/s/www.outlookindia.com/business/germany-commits-10-billion-euros-to-india-for-climate-action-news-194668/amp
"Global Covenant of Mayors Launches Regional Chapter in South Asia," *International Urban Cooperation, European Union-South Asia Chapter.* https://iuc.eu/news/gcom-launch/
"Govt. Driving Transition from Linear to Circular Economy," *PIB, NITI Aayog*, 18 March 2021. https://pib.gov.in/PressReleasePage.aspx?PRID=1705772
"Green Strategic Partnership," Official Website of the Ministry of Foreign Affairs of Denmark, 2021. https://indien.um.dk/en/denmark-in-india/green-strategic-partnership
"Greenko, Italy's Snam to Collaborate on Green Hydrogen," *The Hindu*, 6 November 2020. https://www.google.com/amp/s/www.thehindu.com/news/cities/Hyderabad/greenko-italys-snam-to-collaborate-on-green-hydrogen/article33043387.ece/amp/
Gupta, Uma. "ISA, UNDP Launch Pilot Projects on Farmland Solar in Ten African Member Countries," *PV Magazine,* 27 March 2023. https://www.pv-magazine-india.com/2023/03/27/isa-undp-launch-pilot-projects-on-farmland-solar-in-ten-african-member-countries/
Harvey, Fiona and Suzanne Goldenberg. "The Key Players at the Paris Climate Summit," *The Guardian*, 7 December 2015. https://www.theguardian.com/environment/2015/dec/07/paris-climate-summit-key-players
Hristova, Anna and Dobrinka Chankova. "Climate Diplomacy—A Growing Foreign Policy Challenge," *Juridical Tribune*, 10(2), 2020, pp. 194–206. https://oaji.net/pdf.html?n=2021/5275-1617447586.

Hurrell, A. & Sengupta, S. "Emerging Powers, North-South Relations and Global Climate Politics," *International Affairs*, 88(3), 2012, pp. 463–484.

"India and US Launch the Climate Action and Finance Mobilization Dialogue (CAFMD)," PIB, Ministry of Environment, Forest and Climate Change, 13 September 2021. https://pib.gov.in/PressReleasePage.aspx?PRID=1754590

"India Can Achieve Energy Independence by 2047: US Study," *India Today*, 20 March 2023. https://www.google.com/amp/s/www.indiatoday.in/amp/world/story/india-can-achieve-energy-independence-by-2047-us-study-2348892-2023-03-20

"India, EU to Step up Cooperation on Clean Energy and Climate Action," *The Print*, 8 September 2022. https://www.google.com/amp/s/theprint.in/world/india-eu-to-step-up-cooperation-on-clean-energy-and-climate-action/1121112/%3famp

"India Fastest in Renewal Energy Capacity Addition Among Major Economies," *Economic Times Energyworld*, 10 March 2023. https://www.google.com/amp/s/energy.economictimes.indiatimes.com/amp/news/renewable/india-fastest-in-renewal-energy-capacity-addition-among-major-economies/98530834

"India Has Been Ranked Third Largest Primary Energy Consumer in the World," *PIB Ministry of Petroleum and Natural Gas, Govt. of India*, 24 March 2022. https://pib.gov.in/PressReleaseIframePage.aspx?PRID=1809204

"India Jumps 6 Places to 37th Rank on IMD's World Competitiveness Index," *The Hindu*, 15 June 2022. https://www.google.com/amp/s/www.thehindu.com/news/national/india-jumps-6-places-to-37th-rank-on-imds-world-competitiveness-index/article65529307.ece/amp/

"India's Resolve for Climate Commitments Evident from Performance: PM to G7," *Press Trust of India (PTI)*, *Business Standard*, 27 June 2022. https://www.google.com/amp/s/wap.business-standard.com/article-amp/economy-policy/india-s-resolve-for-climate-commitments-evident-from-performance-pm-to-g7-122062700896_1.html

"India Retains 3rd Position in Renewable Energy Investment Attractiveness Index," *Livemint*, 13 August 2021. https://www.google.com/amp/s/www.livemint.com/news/india/india-retains-3rd-position-in-renewable-energy-investment-attractiveness-index/amp-11634116311363.html

"India, US, Brazil to Work Together for Global Biofuels Alliance," *ET Energyworld*, 2023. https://energy.economictimes.indiatimes.com/amp/news/oil-and-gas/india-us-brazil-to-work-together-for-global-biofuels-alliance/97833089

"Intergovernmental Panel on Climate Change, Climate Change 2022: Impacts, Adaptation, and Vulnerability," *Contribution of Working Group II to the Sixth Assessment Report of the Intergovernmental Panel on Climate Change, Geneva, Intergovernmental Panel on Climate Change*, 2022. https://www.ipcc.ch/report/ar6/wg2/

"International Solar Alliance and West African Power Pool Hosts 13 African Countries in New Delhi to Share Best Practices in Solar Deployment," *PIB*, 14 February 2023. https://www.pib.gov.in/PressReleasePage.aspx?PRID=1899198

Jaspal, Mannat. "Green Multilateralism: Partnerships, Finance, and Innovation: 2021 India-Germany-EU Dialogue," *ORF*, 17 February 2022. https://www.orfonline.org/research/green-multilateralism-partnerships-finance-and-innovation/?amp

Joshi, Madhura. "Strengthening Climate Diplomacy: An Imperative for Indian Climate in the New Decade," *Observer Research Foundation (ORF)*, 26 April 2021. https://www.google.com/amp/s/www.orfonline.org/expert-speak/strengthen-ing-climate-diplomacy-imperative-indian-climate-new-decade/

Kala, Rishi Ranjan. "Quad Eyes India as Manufacturing Hub for Green Hydrogen," *The Hindu Business Line*, 28 June 2022. https://www.google.com/amp/s/www.the-hindubusinessline.com/markets/commodities/quadeyes-india-as-manufacturing-hub-for-green-hydrogen/article65576219.ece/amp/

Kaul, Aastha, "The India-Africa Partnership for Sustainability," *ORF Special Report No. 88*, 7 April 2019. https://www.google.com/amp/s/www.orfonline.org/research/special-report-the-india-and-africa-partnership-for-sustainability-49334/

Koshy, Jacob. "G7 Ministers Commit to Move to Carbon-free Power by 2035," *The Hindu*, 16 April 2023. https://www.thehindu.com/news/international/g7-vows-to-step-up-moves-to-renewable-energy-zero-carbon/article66743606.ece

Kumar, Arun M. "How the Quad Can Help Climate Action," *The Indian Express*, 22 January 2022. https://www.google.com/amp/s/indianexpress.com/article/opinion/columns/how-the-quad-can-help-climate-action-7736009/lite/

Kumar, Manish. "Union Budget Gives India's Green Hydrogen Mission a Shot in the Arm," *Mongabay*, 9 February 2023. https://india.mongabay.com/2023/02/union-budget-gives-indias-green-hydrogen-mission-a-shot-in-the-arm/

Laskar, Rezaul. "Germany Announces €1 Billion Funding for Energy Transition Projects in India," *Hindustan Times*, 30 November 2022. https://www.google.com/amp/s/www.hindustantimes.com/india-news/germany-announces-1-billion-funding-for-energy-transition-projects-in-india-101669827096447-amp.html

Lesolle, David. "The Group of 77 and China's Participation in Climate Change Negotiations," Heinrich Böll Stiftung, 3 February 2014. https://za.boell.org/en/2014/02/03/group-77-and-chinas-participation-climate change-negotiations-publications

"List of SIDS," *UN Office of the High Representative for the Least Developed Countries, Landlocked Developing Countries and Small Island Developing States.* https://www.un.org/ohrlls/content/list-sids

Manchanda, Harish Kumar. "Why Natural Gas Is a Strategic Fuel in India's Energy Transition," *Energyworld, The Economic Times*, 13 December 2022. https://www.google.com/amp/s/energy.economictimes.indiatimes.com/amp/news/oil-and-gas/why-natural-gas-is-a-strategic-fuel-in-indias-energy-transition/96196767

Mohan, Anirudh. "From Rio to Paris: India in Global Climate Politics," *Observer Research Foundation* (ORF), 14 December 2017. ISBN: 978-93-87407-33-6. https://www.orfonline.org/research/rio-to-paris-india-global-climate-politics/?amp

Mohapatra, Satyen. "CoP 27: How Developed World Has Taken the Lead for Climate Change," *The Week*, 4 December 2022. https://www.google.com/amp/s/www.theweek.in/theweek/current/2022/11/25/climate-change-responsibility-taken-by-the-world-at-COP27.amp.html

Neack, Laura. "Middle Powers Once Removed: The Diminished Global Role of Middle Powers and American Grand Strategy," Conference Proceedings.

Columbia International Affairs Online (CIAO), 2000. https://web.archive.org/web/20020228031316/http://www.ciaonet.org/isa/nal01/

Oguntuase, Oluwaseun. "A Climate Emergency: What the IPCC's 2021 Report Means for Africa," The Republic, 16 August 2021. https://republic.com.ng/august-september-2021/a-climate-emergency/

Oguntuase, Oluwaseun J. "India and Africa Leverage Climate Diplomacy," ORF Issue Brief No. 628, Observer Research Foundation, 31 March 2023. https://www.google.com/amp/s/www.orfonline.org/research/india-and-africa-leverage-climate-diplomacy/

"Outcome of COP25 Balanced, with the Exception of Climate Finance Issues: Shri Prakash Javadekar," *PIB, MoEFCC*, 20 December 2019. https://pib.gov.in/PressReleasePage.aspx?PRID=1597047

"PFC Issues India's first-ever Euro Green Bond," *PIB, Ministry of Power*, 16 September 2021. https://pib.gov.in/PressReleaseIframePage.aspx?PRID=1755380

"Quad Leaders' Joint Statement," *White House*, 24 May 2022. https://www.whitehouse.gov/briefing-room/statements-releases/2022/05/24/quad-joint-leaders-statement/

"Renewable Energy in India," *PIB, MNRE, Govt. Of India*, 9 September 2022. https://pib.gov.in/FeaturesDeatils.aspx?NoteId=151141&ModuleId%20=%202

"Report: COVID-19 Slows Progress Towards Universal Energy Access," Departmental News, *World Health Organisation, Geneva, Switzerland*, 1 June 2022. https://www.who.int/news/item/01-06-2022-report--covid-19-slows-progress-towards-universal-energy-access

Report of the Conference of the Parties on Its Twenty-First Session, Held in Paris from 30 November to 13 December 2015, *UNFCCC, FCCC/CP/2015/10/Add.*22, 9 January 2016, p. 2. https://unfccc.int/resource/docs/2015/cop21/eng/10a02.pdf#page=2

Roche, Elizabeth. "EU Joins India's Disaster Resilient Infrastructure Initiative," *Livemint*, 18 March 2021. https://www.google.com/amp/s/www.livemint.com/news/world/eu-joins-india-s-disaster-resilient-infrastructure-initiative/amp-11616062901369.html

Rodrigues, Savio. "India, Saudi Arabia Tango to 'Renewable Energy,'" *Goa Chronicle*, 12 January 2023. https://goachronicle.com/india-saudi-arabia-tango-to-renewable-energy/

Roy Chaudhury, Dipanjan. "European Investment Bank Joins India-Led Coalition for Disaster Resilient Infrastructure," *Economic Times*, 19 July 2022. https://www.google.com/amp/s/m.economictimes.com/news/india/european-investment-bank-joins-india-led-coalition-for-disaster-resilient-infrastructure/amp_articleshow/92988083.cms

Roy Chaudhury, Dipanjan. "India-Italy Partnership: Energy Transition and Circular Economy," *Economic Times*, 12 March 2022. https://www.google.com/amp/s/m.economictimes.com/news/india/india-italy-partnership-energy-transition-and-circular-economy/amp_articleshow/90173733.cms

Saxena, Ankit. "India's CDRI Seeks to Expand Membership to Africa to Create Sustainable and Disaster Resilient Infrastructure," *Swarajya*, 10 April 2023.

https://swarajyamag.com/amp/story/infrastructure%2Findias-cdri-seeks-to-expand-membership-to-africa-to-create-sustainable-and-disaster-resilient-infrastructure

Shaji, S. and Susarla, Arvind. "India, Africa, and Climate Diplomacy," *The Indian Express*, 14 February 2021. https://indianexpress.com/article/opinion/india-africa-and-climate-diplomacy-7188554/

Singh, Siddharth Ghanshyam and Kaifi Jawed. "GOBAR-Dhan: Scheme Announced in Budget a Welcome Step, but Challenges Ahead," *Down to Earth*, 8 February 2023. https://www.downtoearth.org.in/blog/energy/gobar-dhan-scheme-announced-in-budget-a-welcome-step-but-challenges-ahead-87558

Sinha, Sumant. "Why India Is the New Hotspot for Renewable Energy Investors," *World Economic Forum*, 14 January 2020. https://www.weforum.org/agenda/2020/01/india-new-hotspot-renewable-energy-investors/

"Speech by President von der Leyen at the International Solar Alliance," *European Commission*, 24 April 2022. https://ec.europa.eu/commission/presscorner/detail/en/speech_22_2602

Srivastava, Rhea. "Cities as Global Actors: Enhancing Urban Leadership at G20," *Economic Times*, 30 January 2023. https://www.google.com/amp/s/m.economic-times.com/news/international/world-news/cities-as-global-actors-enhancing-urban-leadership-at-g20/amp_articleshow/97436720.cms

Thaker, Jagdish and Leiserowitz, Anthony, "Shifting Discourses of Climate Change in India." *Springer Publications*, DOI 10.1007/s10584-014-1059-6, p. 1, 2014.

Thakker, Aman Y. "India at the United Nations: An Analysis of Indian Multilateral Strategies on International Security and Development," ORF Occasional Paper No. 148, March 2018. https://www.google.com/amp/s/www.orfonline.org/research/india-at-the-united-nations-an-analysis-of-indian-multilateral-strategies-on-international-security-and-development/

Trivedi, Vinay and Shobhit Srivastava, "Clean Push: Why Compressed Biogas Has an Edge Over CNG," *Down to Earth*, 20 January 2021. https://www.google.com/amp/s/www.downtoearth.org.in/news/energy/amp/clean-push-why-compressed-biogas-has-an-edge-over-cng-74874

"UAE Willing to Contribute to India's 500-GW Clean Energy Goal: COP28 President-Designate Al Jaber," *Economic Times*, 22 February 2023. https://www.google.com/amp/s/m.economictimes.com/news/international/uae/uae-eyes-renewables-partnerships-with-india-jaber-says/amp_articleshow/98152756.cms

"U.S. India Strategic Clean Energy Partnership Ministerial Joint Statement," *Press Information Bureau (PIB), Ministry of Petroleum and Natural Gas, Govt. of India*, 7 October 2022. https://pib.gov.in/PressReleasePage.aspx?PRID=1865953

Vachhatani, Jitesh. "MEA Highlights PM Modi's '5S Vision' at UNSC after Being Elected, Outlines 4-Point Mission," *Republic World*, 18 July 2020. https://www.google.com/amp/s/www.republicworld.com/amp/india-news/general-news/mea-highlights-pm-modis-5s-vision-at-unsc-after-being-elected-outl.html

Vestre, Jan Christian and Ville Skinnari. "A Nordic-India Connect to Power a Green Transition," *The Hindu*, 9 February 2023. https://www.google.com/amp/s/www.thehindu.com/opinion/op-ed/a-nordic-india-connect-to-power-a-green-transition/article66486434.ece/amp/

Viswanathan, H.H.S. and Abhishek Mishra. "The Ten Guiding Principles for India-Africa Engagement: Finding Coherence in India's Africa Policy," *ORF Occasional Paper*, 25 June 2019. https://www.google.com/amp/s/www.orfonline.org/research/the-ten-guiding-principles-for-india-africa-engagement-finding-coherence-in-indias-africa-policy/

"Voice of Global South Summit," *Press Information Bureau (PIB), Ministry of Petroleum and Natural Gas, Govt. of India*, 13 January 2023. https://pib.gov.in/PressReleaseIframePage.aspx?PRID=1891153

"What Is the European Global Gateway Strategy?" *French Foreign Policy, Ministry for Europe and Foreign Affairs.* https://www.diplomatie.gouv.fr/en/french-foreign-policy/development-assistance/news/2023/article/what-is-the-european-global-gateway-strategy

Wyns, Arthur. "CoP 27 Establishes Loss and Damage Fund to Respond to Human Cost of Climate Change," *The Lancet Planetary Health*, 7(1), E21–22, 8 December 2022. https://doi.org/10.1016/S2542-5196(22)00331-

"Year—End Review 2022—Ministry of New and Renewable Energy," *PIB*, *MNRE*, 20 December 2022. https://pib.gov.in/PressReleasePage.aspx?PRID=1885147

Chapter Nine

Climate Change Education and the Sustainable Development Goals in Nigeria

Edidiong Samuel Akpabio, Ifeanyi Emmanuel Uzochukwu, Chinonye Frances Kamsi Dike, and Ogunbadeniyi Adeola Oluwakemi

One of the most significant issues of concern in the contemporary world today can be referred to as climate change. The reason for this allusion is not far-fetched, as studies have shown that climate challenges are surging across the globe (Franco et al., 2020), constituting a threat to the sustainable development goals. To demonstrate the volatility of climate change to humanity, it has been described as a risk multiplier that exacerbates humanity's challenges and increases the vulnerability of communities and systems (UNICEF, 2019). In fact, the issues surrounding climate change are clear indicators that it continues to act as a barrier to achieving the SDGs by posing challenges on various fronts (Koubi, 2019). Little wonder why countries have been urged to implement systems that will conserve the environment and factor in its contribution to socio-economic development (UN Environment & ADB, 2019).

It is disturbing to observe that the anomalous situation of climate change, which has defied myriad strategies adopted to confront it, has led to a plethora of challenges for humanity and the environment. Therefore, the development of climate change education and education for sustainable development serves as a means of sensitizing the populace to the requisite steps needed to combat this menace (International Alliance of Leading Educational Institutes, 2009). Furthermore, it is worrisome to note that despite numerous conferences and international engagements by scientists, scholars, and policymakers aimed at raising awareness about the adverse impacts of climate change if left unchecked, the menace seems to continue unabated. This implies that there is a gap that must be filled, and climate change education represents a vital component in addressing this issue. As Lorenzoni & Pidgeon point out, it remains a matter of grave concern that in an era of heightened climatic dysfunctionality, a large proportion of the population seems to lack the requisite knowledge of what climate change entails (Lorenzoni & Pidgeon, 2006). This underscores why Impact Hub describes climate change education as a necessity for a world plagued by the adverse consequences of climate change, as it plays a central role in increasing people's sensitivity to climate change (Impact Hub, 2022).

Granted that climate change continues to pose a significant threat to humanity, laden with catastrophic consequences. This universal threat is believed to have more vulnerability in the developing world due to the state of preparedness in such regions. It is equally not news to state that the African continent is highly vulnerable to climate change despite not being a major contributor to climate variability. While this abnormality is evident, the reason for it can be attributed to the seemingly low level of climate change education in the region. For a country like Nigeria, climate change education requires a sense of urgency and imminence to achieve significant outcomes, especially with respect to SDG attainment. Equally important to emphasize are the adverse implications of climate change on the SDGs for developing countries such as Nigeria, which missed the Millennium Development Goals and may face a repeat scenario with the SDGs if requisite steps are not taken to address this issue promptly.

Research has shown that Sub-Saharan Africa, of which Nigeria is a significant part, has the least capacity to adapt to the fallout of climate change (Rumble & Gilder, 2022). This limited capacity represents a gap that could be addressed through increased investments in climate change education. Considering the benefits that could result from such education, including the cultivation of a new generation of change agents, it is undeniable that climate change education is essential for effectively combating climate change (UNESCO, 2022). This assertion is supported by UNESCO's position, which emphasizes the importance of ensuring that every school is prepared for the challenges posed by climate change in order to advance the Sustainable Development Goals (UNESCO, 2019).

In alluding to this position, Tunji-Olayemi et al. posit that climate change education can help reduce the consequences of climate change on humanity and the environment (Tunji-Olayeni et al., 2021). They argue that it is an activity that elicits action (Sibanda & Manik, 2023) and represents a response mechanism to the adverse effects of climate change (IPCC, 2018). This aligns with the viewpoint of Fahey et al., who affirm that training individuals in climate change education can enable them to contribute to the Sustainable

Development Goals (Fahey et al., 2016), particularly considering that climate change education utilizes educational platforms to achieve the goals embedded in SDG 13 (Osama et al., 2018). Previous studies have revealed the critical role of education in the SDG agenda (Mochizuki & Bryan, 2015). According to Selby (2007), the importance of educating youths on the dangers of climate change cannot be overemphasized. This provides justification for why SDG 13 adopts climate change education as a platform to combat climate change (Glavic, 2020).

The crux of this work is therefore hinged on appraising the state of the SDGs in Nigeria while establishing a nexus between climate change education and the global goals. The aim is to unearth problems that can be fixed and prospects that can be leveraged upon to ensure that the SDGs become a reality in Nigeria. This is premised on the understanding that while SDG 4 explicitly addresses education, it is foundational to most of the SDG goals. We posit that climate change education represents a veritable platform for actualizing the SDGs. The chapter asserts that climate change education is laden with possibilities and potentials, although there are embedded problems and perils associated with it. Preliminary studies have revealed a knowledge gap in Nigeria, with climate change illiteracy levels high because the majority of the populace are unaware of the effects of climate change. Therefore, we advocate for compulsory climate change education at all stages of education in Nigeria, as climatic reforms can only happen when the people are sufficiently informed.

CONCEPTUAL DISCOURSES

Climate change, apart from threatening the individual SDG goals, is capable of stalling overall attainment, hence the need to implement measures such as climate change education to mitigate its continuous spread. The concept of climate change education is simply a response to the longstanding challenge of climate change. It seeks to enlighten students on how they can address the adverse state of the climate through engaging in mitigative and adaptive practices. It is widely acknowledged that human activities have contributed to climate change (as alluded to by the Intergovernmental Panel on Climate Change, 2007, and the National Research Council, 2010). Therefore, human activities can also significantly contribute to reducing climate change. This is why it is posited that one-way human activities can mitigate the damage caused by climate change is through their involvement in climate change education. To demonstrate the utility and indispensability of everyone in the climate change education chain, Cantell et al. (2019) appraised climate change education through the lens of the bicycle model. The use of this model by the authors aims to depict effective participation by all in the fight against climatic dysfunctionality. Demaidi & Al-Sahili affirmed that schools especially universities play a key role in promoting climate change awareness (Demaidi & Al-Sahili, 2021), hence, it remains a truism that for climate change awareness to be effective, universities will play a key role in making this happen (Demaidi & Al-Sahili, 2021). We must re-emphasize that for climate change education to be productive in Nigeria and deliver on the requisite expectations, it must be integrated into the mainstream of the nation's educational spectrum. While it might look worthwhile to suggest that climate change education be integrated into the school curriculum at every stratum of the educational ladder, the question to ask is if the teachers themselves have acquired sufficient training on climate change to be in good stead to disseminate valuable knowledge to their students.

While climate change education represents a response to the menace of climate change, Siegner & Stapert observed that disseminating climate change education has proven to be difficult to achieve in schools, even beyond this region (Siegner & Stapert, 2019). This challenging state of affairs can be attributed to noticeable gaps observed in climate change education (Vaughter, 2016). Research has shown that the fundamental objective of climate change education is to achieve a shift in behavioral patterns and attitudinal dispositions (Boakye, 2015). These shifts, targeted at reducing climate change, can be achieved through the intensification of climate change education, with the primary objective of influencing behavioral modifications. Studies have also shown that education has the capacity to empower individuals, as it can motivate young people to take action. Hence, there is a need to integrate climate change education into classrooms (United Nations, 2022b). According to Molthan-Hill and co-researchers, if every student has basic literacy on climate change adaptation and mitigation, climate change can be significantly controlled and reduced in the long run. This explains why climate change education has been proposed for integration at institutional, national, and international levels (Molthan-Hill et al., 2022). It is expedient to reiterate that for climate change education to be beneficial, successful, and rewarding, it must be holistic and all-encompassing. It should transcend the confines of classrooms, and teaching methodologies should be enhanced to include more practical sessions. In Nigeria, there are gaps in climate change education, as even most conventional schools have failed to integrate CCE into their curriculum. This is a major reason why significant strides have not been achieved in climate change mitigation in Nigeria. According to UNESCO, education is crucial to promote climate action (UNESCO, 2023). Unfortunately, Africa's average national climate change literacy is pegged at 37%, contrasting sharply with the over 80% recorded across North America and Europe. This underscores the urgent need for an emergency response to climate change education in Africa (Simpson et al., 2021). It is equally important to emphasize at this juncture that climate change education can be delivered in both formal and informal settings. In other words, CCE is not bound by a rigid method of dissemination. Thus, teaching climate change education can utilize a variety of strategies. For instance, Abaraonye et al. (2018) suggested the utilization of poetry as a medium for climate change education, a proposal supported by Lilenfield et al. (2018), who advocated for the integration of nursing as a platform to address climatic dysfunctions. Regardless of the medium utilized, Boxer (2021) argues that climate-related issues should hold more relevance over every other subject being taught.

CLIMATE CHANGE AND THE SDGS IN NIGERIA

Climate change events represent critical developmental issues for the Nigerian state, considering its degree of vulnerability. The Federal Government of Nigeria (FGN) recognizes climate change as a serious threat to its sustainable growth in the decades to come, with the potential to cause damaging and irrecoverable effects on infrastructure, food production, water supplies, revenue generation, and precipitating natural resources conflicts (FMEN, 2021). Without proper adaptive and mitigation strategies against climate change scenarios, Nigeria could experience a GDP loss of about 6% to 30% (amounting to about US$ 100 billion to 460 billion) by 2050 (DFID, 2009; Ebele et al., 2016). On the other hand, the country's total greenhouse gas emissions (GHG), although currently low (less than 1% of the global value), are expected to significantly increase in the near future given the projected rapid economic growth driven by the need to meet the demands of its rapidly increasing human population (FMEN, 2021). Therefore, to make significant progress in addressing global climate change issues, there is a need for a holistic transformation of the world's economies and societies. Nigeria especially needs to direct efforts towards reducing or regulating its global contribution of GHGs, but more importantly, towards increasing its preparedness and adaptability to future climate change dynamics. The United Nations (United Nations, 2022a) has reiterated the need to make changes in various sectors such as energy, industry, agriculture, and transport, as well as at institutional and individual levels, to reduce their emissions while adapting to contemporary climate change events. Therefore, analyzing the relationship between these sectors and climate change concerns is critical for attaining sustainable development and the actualization of the various SDGs.

Furthermore, the existence of interlinkages between the SDGs, which have been described as integrated and indivisible by Tosun & Leininger (2017), implies that careful consideration must also be given to the synergies and trade-off relationships among the SDGs for meaningful progress to be made in their attainment. Understanding these interactions is critical for policy coherence and the focused prioritization of action plans to achieve appreciable synergistic impacts (UN-DESA, 2018). While concerns persist regarding the practicability of achieving the set targets of some SDGs in light of the trade-offs between increased output and sustainability (low-carbon approach), efforts to increase the adaptation capacities of various developmental sectors to climate change impacts will certainly enhance the actualization of other goals. Therefore, it is necessary to consider the interlinks between climate change issues and SDG 13 (Climate action), as well as the achievements made in other SDGs, particularly those with strong climate change links. Understanding the present capacity and possible future tendencies of various sectors of the economy for greenhouse gas emissions remains pertinent in appreciating the country's level of entanglements and contribution to climate change, and in forming effective policies to drive low-carbon sustainable economic growth and development. Moreover, although climate change issues affect all critical sectors of society and are intricately interwoven with the other 16 goals of the SDGs, the IPCC has suggested that robust synergies be forged among SDGs 3 (Good health and well-being), 7 (Affordable and clean energy), 11 (Sustainable cities and communities), 12 (Responsible consumption and production), and 14 (Life below water/oceans) for substantial achievements to be made in the proposed climate change targets.

Food and Livelihoods: According to Fada et al. (2022), the interaction between SDG 13 and SDGs 1 (No poverty) and 2 (Zero hunger) is mutually reciprocal. Agriculture, forestry, and other land use (AFOLU) emit approximately 366,734 Gg CO_2-eq, representing about 60% of the total net GHG emissions in the country (FMEN, 2021). However, despite significant efforts by the Federal Government of Nigeria, with the collaboration and support of several international agencies, millions of Nigerians are continuously plunged into severe hunger and poverty crises daily. Recent reports show that about 8.4 million people are food insecure, with 70% of the country's population living below the poverty line, and conflicts in the Northeast region displacing over 3 million people (Global Network Against Food Crisis, 2022; World Food Programme, 2022). The key drivers of this food insecurity crisis include increasing conflicts and insecurity, climate change events such as droughts and floods, and economic shocks, including COVID-19, not excluding dwindling oil production and revenue (Sachs et al., 2022). The overarching influence of weak institutions and political will (Goal 16: Peace, Justice, and Strong Institutions) to squarely engage with these issues continues to hinder the impact of any well-meaning vision, policy, and agreements that the country has adopted. Therefore, addressing this challenge sufficiently and adapting other critical sectors to climate change events would not only ensure improved food production to end hunger but will also contribute to poverty alleviation, especially as Nigeria has about 35% of its population engaged in the agricultural sector (World Bank, 2022).

Health and Well-Being: Health indicators in Nigeria rank among the worst in Africa, with the highest burden of global malaria cases, neonatal mortality rates, and an estimated second-highest number of global maternal deaths and people living with HIV/AIDS (USAID, 2022; World Health Organization, 2015). The direct adverse impacts can be observed in areas such as food and water shortages, which have resulted in massive movements of people and animals, leading to the spread of diseases (Tosun & Leininger, 2017). Additionally, increasing average temperature ranges and other extreme weather events are expected to directly cause heat-related deaths, threaten lives and property with extreme weather events, reduce air quality, and increase the incidence of allergies, respiratory issues, and mental health concerns (Luber et al., 2014). This is compounded by the epidemic potential of tropical diseases like malaria in Nigeria (Fernando, 2022). According to the Global Network Against Food Crisis (2022), about 35 million people in Nigeria are currently stressed, with an additional 12.7 million in crisis. Achieving clean water (SDG 6) will undoubtedly lead to a reduced spread of diseases and improved health (Watts et al., 2018). Additionally, there is an issue of increased migratory patterns (southwards) resulting from increasing climate change incidence in northern Nigeria (Fada et al., 2022). Consequently, there has been heightened pressure on limited resources and increasing conflicts between farmers and herders, resulting in a significant loss of human lives. Formulating policies for sustainable development in the health sector will require a critical assessment of these existing and expected climate change-stimulated scenarios to better adapt health systems to sustainable development.

Energy: The energy sector contributes about 33% (total of 206,452 Gg CO_2-eq) of the gross GHGs emission in the country (Climatelinks, 2019). Besides being the second highest contributor of GHGs in Nigeria, this sector is also critical for its role in the possible transformation and development of other sectors. Also, with about 35 billion barrels of oil reserves, 5.8 trillion m3 of gas reserves, and 21 million tonnes of proven recoverable bituminous coal reserves, Nigeria will undoubtedly be depending on carbon and fossil fuel energy for its economic development at least soon (FMEN, 2021). Hence, envisioning and achieving a low-carbon development for the country will be a challenging one as its mono-economy is hugely dependent on oil and gas. Thus, it could prove difficult to decouple emissions from the country's economic growth. Nigeria is therefore set for a huge unstable and vulnerable fiscal and macroeconomic future, especially with global efforts to reduce GHGs emissions and the highly variable oil prices associated with the global transitioning to green economy. Therefore, addressing the challenges of climate change to national development remains the most critical and fundamental consideration for achieving long-term economic planning effectiveness in Nigeria. Although, it has been argued that this may actually present an important opportunity for the country to diversify its economy and stimulate the development of low-carbon technologies (carbon capture and storage), energy efficiency technologies and fully exploit its large low-carbon alternative energy resources (sun, wind, hydro, biomass, and tidal waves) (FMEN, 2021). Besides, increasing the efficiency of energy use, the utilization of renewable energy (SDG7) is expected to aid the attainment of SDG1 (poverty), SDG2 (hunger), and SDG13 (climate change) (Fada et al., 2022). The removal of subsidies on fossil fuel use will also support the reduction of wasteful production and consumption (SDG12), and by extension, energy use (SDG7) and climate change (SDG13).

Energy: The energy sector contributes approximately 33% (a total of 206,452 Gg CO_2-eq) of the gross GHG emissions in the country (Climatelinks, 2019). Besides being the second-highest contributor of GHGs in Nigeria, this sector is also critical for its role in the possible transformation and development of other sectors. With about 35 billion barrels of oil reserves, 5.8 trillion m3 of gas reserves, and 21 million tonnes of proven recoverable bituminous coal reserves, Nigeria will undoubtedly continue to rely on carbon and fossil fuel energy for its economic development in the near future (FMEN, 2021). Envisioning and achieving a low-carbon development for the country will be challenging, given its heavy dependence on oil and gas and the mono-economy structure. Decoupling emissions from the country's economic growth could prove difficult. Nigeria is therefore facing a potentially unstable and vulnerable fiscal and macroeconomic future, especially with global efforts to reduce GHG emissions and the highly variable oil prices associated with the global transition to a green economy. Addressing the challenges of climate change to national development remains the most critical and fundamental consideration for achieving long-term economic planning effectiveness in Nigeria. Although it has been argued that this may present an important opportunity for the country to diversify its economy and stimulate the development of low-carbon technologies, such as carbon capture and storage, energy efficiency technologies, and fully exploit its large low-carbon alternative energy resources, such as solar, wind, hydro, biomass, and tidal waves (FMEN, 2021). Furthermore, besides increasing the efficiency of energy use, the utilization of renewable energy (SDG 7) is expected to aid in the attainment of SDG 1 (poverty), SDG 2 (hunger), and SDG 13 (climate change) (Fada et al., 2022). The removal of subsidies on fossil fuel use will also support the reduction of wasteful production and consumption (SDG 12), and by extension, energy use (SDG 7) and climate change (SDG 13).

Industry: The significant dependence of the industrial sector on natural gas and diesel for energy generation in Nigeria is a cause for concern due to their GHG emissions (Eleri et al., 2013). The Nationally Determined Contributions (NDC) projected that this sector may increase its GHG emissions from 4.2 Mt CO_2-eq in 2010 to about 14.8 Mt CO_2 in 2030 if adequate measures are not implemented to ensure energy efficiency (FMEN, 2021). However, as of 2018, this has already increased to about 10.35 Mt CO_2-eq (approximately 2.1% of the total) (Climatelinks, 2019), and is expected to continue growing due to the anticipated high rate of industrialization. Of particular concern is the oil and gas sector of the country, where gas flaring, fugitive methane emissions, on-site fuel use (both upstream and downstream), and widespread artisanal crude refining activities (using rudimentary equipment) in the Niger Delta region significantly contribute to GHG emissions and environmental degradation. Building resilient infrastructures, promoting inclusive and sustainable industrialization, and fostering innovation in Nigeria will therefore require a transition in industrial power generation towards the use of natural gas to increase output and reduce GHG emissions.

Urbanization and Waste Management: With the increasing human population, industrialization, urbanization, expansion of existing cities, and economic growth in Nigeria, the volume of municipal solid waste (MSW) is expected to significantly increase (Adeleke et al., 2021). This is unsurprising given Nigeria's status as one of the highest waste generators in Africa, with an annual MSW generation of 32 million tonnes and a formal waste collection rate of 20–40% (Sowunmi, 2019). Consequently, the poor waste management system, coupled with the indiscriminate waste disposal practices of households, markets, and industries, adversely impacts the aesthetics of the environment and contributes significantly to GHG emissions (approximately 68.8 Mt CO_2-eq, accounting for 14% of the total GHG emissions) (Climatelinks, 2019). Adopting effective urban centers and waste management strategies, leveraging the multi-layered benefits of mitigation, adaptation, and improved access to services, will help tackle GHG emissions and enhance climate change (CC) mitigation and the quality of life. Furthermore, efforts focused on exploring and developing technologies for effective waste collection and converting waste to energy (methane gas and electricity) will not only reduce GHG emissions but also contribute to achieving affordable and clean energy (SDG 7), sustainable cities and communities (SDG 11), and responsible consumption and production (SDG 12) in both urban and rural communities.

Marine and Water Resources: There is a synergistic relationship between SDG 13 on climate change action and SDG 14 (Life below water) (Fada et al., 2022). The issues of burning water tributaries in the Niger Delta due to oil spills, along with the activities of wharves, seaports, and marinas, are all known to contribute to GHGs emissions. Conversely, the significant threats of climate change-stimulated droughts, floods, rising ocean temperatures, water acidification, and reductions in water supply are expected to adversely affect agriculture, industries, energy generation, fisheries, biodiversity, and human health. Achieving climate change actions (SDG 13) will improve the health of marine ecosystems and life, enabling the use of these resources to achieve sustainable development (Fada et al., 2022). Coastal ecosystems are vital carbon sinks, and their conservation would enable them to contribute to the mitigation and adaptation to climate change events while strengthening the resilience of coastal communities (Schmidt et al., 2017).

Transportation: One critical sector contributing significantly to GHG emissions in the country is the transportation sector. It accounts for a substantial portion of carbon growth in human population, economic activity, and wealth (Cervigni et al., 2013). Already, several individuals and companies are adopting solar-powered and electric vehicles within the country. While this is a welcome development, efforts should be made by the Federal Government of Nigeria and international organizations to support such productions and their adaptation for mass vehicle use. However, it is also crucial to consider this in relation to its impact on Nigeria's mono-rhythmic economy, which relies solely on oil for its revenue.

REASSESSING THE CLIMATE CHANGE, CLIMATE CHANGE EDUCATION AND DEVELOPMENT NEXUS

Throughout human history, education has remained a catalyst for societal transformation (Chakraborty et al., 2018), nurturing individuals by instilling self-discipline, fostering positive social engagement and knowledge growth, and ultimately shaping compassionate and adept individuals for progress and adjustment within society (Al-hattami and Al-Ahdal, 2014). Recognizing its relevance, the UNFCCC assigns climate change education as the responsibility of Convention Parties. Its goal is to disseminate information for better understanding, adaptation, and empowerment through behavioral change and advocacy (UNCC, 2022).

Formal education offers comprehensive knowledge on the environment, ecological interactions, climate change events and impacts, sustainability, behavioral modifications, adaptation skills and motivation for sustainable development (UNCC, 2022). To buttress its utilitarian functionality, education has been referred to as a significant social factor in societal advancement, since development involves multiple aspects, including environmental, economic, social and political factors (CDG, 2023). This explains why the importance of climate literacy in attaining sustainable development goals cannot be overstated.

Studies have shown that developing countries lack requisite resources to deal with climate change induced problems. This challenge is not restricted to them as their developed counterparts share similar experiences with the most vulnerable communities inhabiting the areas more prone to extreme climatic events (McCarthy, 2020). This degree of proximity makes them suffer the most sordid impacts which includes loss of employment, displacement among other negatives (Khoday and Ali, 2018).This aside, the predictions of increased poverty in staggering numbers by 2030 due to climate change raises concern (UNDP, 2020), and is poised to increase the number of people living below the poverty line by over 43 million in the African continent (World Bank, 2022).

Furthermore, changing weather patterns, climate hazards such as increased temperature, drought, famine, flooding and increased pest infestation including disease prevalence have grossly lowered agricultural activities, food production and increased cost of food (World Bank, 2022), while productivity of workers is significantly reduced by the communities and countries. Nonetheless, heat stress (ILO, 2019), which inadvertently negatively impacts nutrition especially for poor agricultural emissions (Methane, Nitrous oxide) and unsustainable food production practices contribute immensely to climate change (Lynch et al., 2021).

Climate change not only leads to climate-related diseases but also exacerbates existing health risks (Nolte et al., 2018). The WHO predicts a potential annual increase of 250,000 climate-related deaths by 2030 if global remediation efforts are ineffective (WHO, 2021). Additionally, the rise in climate-related diseases contributes to increased absenteeism and reduced education quality for affected students (UNDP, 2017). Moreover, extreme climate events result in the destruction of social and educational infrastructure, school closures and teacher shortages particularly in vulnerable areas, widening knowledge gaps and increasing the number of out-of-school children especially in the global south (UNDP, 2017). Consequently, Gender inequality worsens particularly in developing countries (Ngepah and Mwiinga, 2022), evidenced by absenteeism from work and school especially for girls and young women within the childbearing age with a concomitant increase in child marriages and intensified child labor which are survival strategies for low-income families (UNDP, 2017). Socio-cultural practices favoring male children in education also come into play when vulnerable families face hardships imposed by climate change.

The evidence of climate change can all be categorized into various types of aquatic catastrophes (United Nations, 2022b). These effects are regional, with some regions experiencing drought and the shrinking of aquatic water bodies, leading to water scarcity, while others face flooding and the submerging of coastal communities due to the rise in sea levels (United Nations, 2023). This results in contamination of fresh water with salt water, water pollution, sanitation deficiencies, and secondary consequences including waterborne diseases, loss of habitats, destruction of infrastructure, and reduced food production.

The energy sector is implicated as the highest contributor to greenhouse gases and a causative factor in climate change (World Bank, 2022**[AQ10]**; EPA, 2023). As the effects of climate change persist, the use of temperature-moderating appliances increases, leading to more greenhouse gas emissions (Zamuda et al., 2018). This phenomenon presents a conundrum. Furthermore, climate hazards can overload power lines, reduce carrying capacity, damage transmission lines, and negatively affect businesses, healthcare, education, and security (EPA, 2023), endangering people's lives (Sathaye et al., 2011). Similarly, the by-products of industrialization, which began during the Industrial Revolution, are the genesis of climate change (Wandana et al., 2020). As industrialization and urbanization advance, the contribution to climate change through the release of greenhouse gases increases. The inequality between countries is aggravated as climate change persists; developed countries historically account for about 79% of carbon emissions (CDG, 2023), but the impact is greater on developing countries. For instance, projected job losses will affect Southern Asia and West African regions the most, with an estimated 9 million jobs lost, equivalent to US$43 million, even though these countries contribute less than 1% to global warming. Countries in the global South, facing socio-economic and political issues, lack the resources to aid adaptation (ILO, 2019), deepening the impact of climate change and exposing deficiencies in their adaptive ability and resilience (African Development Bank, 2019), stifling economic growth. The International Labour Organization (ILO), based on a forecasted global temperature rise of 1.5°C, reports that the global economy will incur a loss of about US$2,400 billion, resulting from a loss of 80 million full-time jobs by 2030 due to global warming-induced heat stress (ILO, 2019). Considering this, the escalation of regional conflicts and terrorism is driven by a rise in climate refugees, who become agitators against the government and susceptible targets for terrorist recruitment (Strozier and Berkell, 2014; Nett and Rüttinger, 2016; Ugwuede and Dike, 2021).

Additionally, limited environmental resources are exploited as a tool for population manipulation (Kohler et al., 2019; Sutter, 2018; Gleick, 2014). Sustainability of cities hinges on utilizing renewable resources and adopting adequate planning, relevant legislation, effective systems, implementation strategies, and climate-literate inhabitants (Berger, 2014). Conversely, unsustainable cities suffer from overpopulation, rely on fossil fuel combustion, lack natural vegetation and green infrastructure, resulting in high carbon emissions. Unsustainable consumption and production patterns exacerbate ecological degradation and contribute to climate change (Arora and Mishra, 2023). The World Economic Forum (2020) highlights the loss of 50% of flora species in terrestrial ecosystems due to deforestation, negative agricultural and urbanization practices. Moreover, municipal food waste contributes around 9% of GHGs (FAO, 2023), contributing to an annual estimate of 3.3 billion tons of carbon dioxide (UNEP, 2021). Sadly, coastal cities, especially in developing countries, are particularly vulnerable to climate change impacts due to a lack of inclusivity and adaptation plans (El-Siofi, 2010).

STRATEGIES TO IMPROVE CLIMATE CHANGE EDUCATION FOR SUSTAINABLE DEVELOPMENT

Climate change is intricately linked to each of the sustainable development goals, and global climate literacy plays a crucial role in finding solutions and fostering innovations to achieve them by 2030. Although climate change education has gained momentum worldwide, disparities exist in its implementation both between and within countries, presenting challenges like non-inclusive education, absence of a curriculum, and lack of collaboration. Therefore, to address these issues effectively, developing countries should consider adoption and implementation of the following strategies:

- Designing a climate change-focused curriculum tailored to suit the unique challenges of each country and region aligning with the SDGs.
- Introducing activity based climate change education at all educational levels to enhance engagement and understanding.
- Provision of comprehensive training for teachers and instructors to emphasize the interconnections of climate change with other concepts.
- Developing inclusive learning materials in multiple languages that cater to diverse learners.
- Empowering grassroots efforts by collaborating with climate change and sustainable development officers to educate communities through town hall gatherings.
- Incorporating climate change lessons into the certification process for business registrations.
- Embracing non-traditional learning methods and utilizing technology for diverse and accessible content delivery.

RECOMMENDATIONS

With the failure of most African states to achieve the millennium development goals and the due date for the sustainable development goals drawing nigh, there is the need to develop an integrated approach to the SDGs as climate change considerations remain centripetal to the achievements of other SDGs. The establishment of the Nationally Determined Contributions (NDCs) and the development of the Long-Term visions for communicating the Long-Term Low Greenhouse Gas Emissions Development Strategies (LT-LEDS), as stipulated in the Paris Agreement, are intended to assist various participating countries to design, redirect and explore new ways of achieving their CC control ambitions. While climate change issues should be given priority, policies and plans must be directed toward an integrated approach to SDGs' achievements (Fada et al., 2022). Additionally, while equal opportunities for education (SDG4) is unarguably, required for the wide understanding of the impacts, risks, and adaptation means to climate change issues, and inadvertently, the achievement of the SDGs (Fada et al., 2022), its attainment remains a critical challenge in Nigeria and the world (United Nations Educational, 2019). Nigeria's education presently receives about 5.4% of the annual budget against the recommendation of the United Nations Scientific and Cultural Organization (UNESCO) that developing nations allot up to 15–20% (UNESCO, 2022). A massive investment in the education sector of the country will undoubtedly, go a long way to increase its achievements in other critical areas of development such as poverty, food security, health, and decent work and economic growth. Furthermore, as suggested by (Akinbami & Akinbami, 2017), directing efforts at researches can also offer practical support to formulation and management of climate policies while exposing challenges and missing gaps for future research. Finally, to make any organic and sustainable development in Nigeria with reference to the SDGs, there is the need for the creation of strong institutions which will ensure the effectiveness and consistency of policy implementations and assessments.

CONCLUSION

This chapter set out to appraise the state of climate change education in Nigeria with a view to establish a nexus between the concept and the sustainable development goals. In a bid to achieve this objective we appraised the state of SDGs in the country which was quite dismal with a cacophony of loopholes precipitated by the climate change debacle which re-emphasizes the fact that the need to develop effective strategies to communicate climate change education cannot be over-emphasized. We observed from the literature scrutinized that the untoward effects of climate change on the Nigerian state requires a sense of urgency in response which climate change education offers. We posit that while climate change is a reality and a crisis, climate change education is a necessity and a mandatory practice to effectively respond to the environmental needs of the contemporary world especially in developing societies like Nigeria. In conclusion, we make a case for the deployment of requisite political will by the leadership class to bail the Nigerian state out of the current climatic quagmire she has been plunged into using the instrumentality of climate change education as a leeway to achieve the SDGs.

BIBLIOGRAPHY

Adeleke, M. L., Dickson, M., & Ayal, D. (2021). "African Handbook of Climate Change Adaptation." In W. L. Filho, N. Oguge, D. Ayal, L. Adeleke, & I. da Silva (Eds.), African Handbook of Climate Change Adaptation (pp. 789–805). https://doi.org/10.1007/978-3-030-45106-6

African Development Bank. (2019). Analysis of Adaptation Components of Africa's Nationally Determined Contributions (NDCS) (pp. 1–46).

Al-Hattami, A., & Al-Ahdal, A. (2014). "An Insight into the Role of Education as an Agent of Change in the Social Development." Social Development. Retrieved from https://www.researchgate.net/publication/268226453_An_Insight_into_the_Role_of_Education_as_an_Agent_of_Change_in_the_Social_DevelopmentE

Anabaraonye, B., Nji, I. A., & Hope, J. (2018). "Poetry as a Valuable Tool for Climate Change Education for Global Sustainability." International Journal of Scientific & Engineering Research, 9(9), 81–84.

Arora, N. K., & Mishra, I. (2023). "Responsible Consumption and Production: A Roadmap to Sustainable Development." Environmental Sustainability, 6, 1–6. https://doi.org/10.1007/s42398-023-00266-9

Berger, M. (2014). "The Unsustainable City." Sustainability, 6(1), 365–374. doi:10.3390/su6010365

Boakye, C. (2015). "Climate Change Education: The Role of Pre-Tertiary Science Curricula in Ghana." Sage Open, 5(4), 2158244015614611.

Cantell, H., Tolppanen, S., Aarnio-Linnanvuori, E., & Lehtonen, A. (2019). "Bicycle Model on Climate Change Education: Presenting and Evaluating a Model." Environmental Education Research, 25(5), 717–731.

Centre for Global Development. (2023). Developed Countries Are Responsible for 79 Percent of Historical Carbon Emissions. Retrieved from https://www.cgdev.org/media/who-caused-climate-change-historically

Cervigni, R., Valentini, R., & Santini, M. (2013). Toward Climate-Resilient Development in Nigeria. Directions in Development—Countries and Regions. Washington, DC: World Bank. https://doi.org/10.1596/978-0-8213-9923-1

Chakraborty Saikat, Bidisha Chakraborty, Virender Singh Dahiya, & Louie Timajo. (2018). "Education as an Instrument of Social Change and Enhancing Teaching-Learning Process with the Help of Technological Development." Technology Development, pp 1–16.

Climatelinks. (2019). Greenhouse Gas Emissions Factsheet: Nigeria. Global Climate Change. Retrieved from https://www.climatelinks.org/resources/greenhouse-gas-emissions-factsheet-nigeria

Collins, M., Knutti, R., Arblaster, J., Dufresne, J.-L., Fichefet, T., Friedlingstein, P., & Wehner, M. (2013). "Long-Term Climate Change: Projections, Commitments and Irreversibility." In T.F. Stocker, D. Qin, G.-K. Plattner, M. Tignor, S.K. Allen, J. Boschung, A. Nauels, Y. Xia, V. Bex, & P.M. Midgley (Eds.), Climate Change 2013: The Physical Science Basis. Cambridge University Press.

Demaidi, M. N., & Al-Sahili, K. (2021). "Integrating SDGs in Higher Education—Case of Climate Change Awareness and Gender Equality in a Developing Country According to RMEI-TARGET Strategy." Sustainability, 13(6), 3101.

Ebele, N. E., Emodi, N. V., & Pezzoli, A. (2016). "Climate Change and Its Impact in Nigerian Economy." Journal of Scientific Research & Reports, 10(6), 1–13. https://doi.org/10.9734/JSRR/2016/25162

El SioufiI, M. (2010). "Climate Change and Sustainable Cities: Major Challenges Facing Cities and Urban Settlements in the Coming Decades" (pp. 1–12). Retrieved from https://www.fig.net/resources/monthly_articles/2010/june_2010/june_2010_el-sioufi.pdf

Eleri, E. O., Onuvae, P., & Ugwu, O. (2013). Low-Carbon Energy Development in Nigeria: Challenges and Opportunities (SUNGAS project). Retrieved from https://www.iied.org/g03555

Environmental Protection Agency (EPA). (2023). Climate Change Impacts on Energy. Retrieved from https://www.epa.gov/climateimpacts/climate-change-impacts-energy#14foot

Fada, S. J., Akintunde, E. A., Goyol, S. S., Binbol, N. L., Bombom, L., Dabi, D. D., Baklit, G., & Oche, C. Y. (2022). "An Appraisal of Nigeria's Progress in Achieving the SDG-13 Climate Action Goal." Journal of Sustainable Development, 15(2), 66–77. https://doi.org/10.5539/jsd.v15n2p66

Fahey, S., Verstraten, L., & Berry, A. J. (2016). "Education for Sustainable Development: Enhancing Climate Change Adaptation Expertise in Developing Countries." Journal of Education for Sustainable Development, 10(1), 54–67.

Food and Agriculture Organization (FAO). (2023). "Food Wastage: Key Facts and Figures." Retrieved from https://www.fao.org/

Fernando, S. D. (n.d.). "Climate Change and Malaria—A Complex Relationship." United Nations Chronicle; United Nations. Retrieved October 16, 2022, from https://www.un.org/en/chronicle/article/climate-change-and-malaria-complex-relationship

FMEN. (2021). 2050 Long-Term Vision for Nigeria (LTV-2050)—Towards the Development of Nigeria's Long-Term Low Emissions Development Strategy (LT-LEDS).

Franco, I. B., Tapia, R., & Tracey, J. (2020). "SDG 13 Climate Action: Climate Education: Identifying Challenges to Climate Action." In Actioning the Global Goals for Local Impact: Towards Sustainability Science, Policy, Education and Practice (pp. 219–228). New York: Springer, 2020.

Glavič, P. (2020). "Identifying Key Issues of Education for Sustainable Development." Sustainability, 12(16), 6500.

Gleick, P. H. (2014). "Water, Drought, Climate Change, and Conflict in Syria." Weather, Climate, and Society, 6(3), 331–340. doi:10.1175/wcas-d-13-00059.1

Global Network Against Food Crisis. (2022). "2022 Global Report on Food Crises: Joint Analysis for Better Decisions." Retrieved from https://docs.wfp.org/api/documents/WFP-0000138913/download/?_ga=2.85351005.593612021.1666082432-1684028073.1666082432

International Labour Organization (ILO). (2023). "Increase in Heat Stress Predicted to Bring Productivity Loss Equivalent to 80 Million Jobs." Retrieved from https://www.ilo.org/global/about-the-ilo/newsroom/news/WCMS_711917/lang--en/index.html

Intergovernmental Panel on Climate Change (IPCC). (2018). Global Warming of 1.5°C, an IPCC Special Report on the Impacts of Global Warming of 1.5°C Above Pre-Industrial Levels and Related Global Greenhouse Gas Emission Pathways, in the Context of Strengthening the Global Response to the Threat of Climate Change, Sustainable Development, and Efforts to Eradicate Poverty. Retrieved from http://www.ipcc.ch/report/sr15/

Kishan, K., & Ali, W. (2018). "Climate Change and the Rise of Poverty." United Nations Development Programme (UNDP). Retrieved from https://www.undp.org/blog/climate-change-and-rise-poverty

Kohler, C., Dos Santos, C. D., & Bursztyn, M. (2019). "Understanding Environmental Terrorism in Times of Climate Change: Implications for Asylum Seekers in Germany." Research in Globalization, 100006. doi:10.1016/j.resglo.2019.100006

Koubi, V. (2019). "Sustainable Development Impacts of Climate Change and Natural Disaster." *Background Paper Prepared for Sustainable Development Outlook.*

Lilienfeld, E., Nicholas, P. K., Breakey, S., & Corless, I. B. (2018). "Addressing Climate Change Through a Nursing Lens Within the Framework of the United Nations Sustainable Development Goals." Nursing Outlook, 66(5), 482–494.

Lorenzoni, I., & Pidgeon, N. F. (2006). "Public Views on Climate Change: European and USA Perspectives." *Climatic Change*, *77*(1), 73–95.

Lynch, J., Cain, M., Frame, D., & Pierrehumbert, R. (2021). Front. Sustain. Food Syst., Sec. Climate-Smart Food Systems, Volume 4–2020. doi:10.3389/fsufs.2020.518039

McCarthy, J. (2020). "Why Climate Change and Poverty Are Inextricably Linked." Global Citizen. Retrieved from https://www.globalcitizen.org/en/content/climate-change-is-connected-to-poverty/#:~:text=An%20estimated%20100%20million%20people,them%20afterwards%2C%E2%80%9D%20Levine%20said.

Mochizuki, Y., & Bryan, A. (2015). "Climate Change Education in the Context of Education for Sustainable Development: Rationale and Principles." Journal of Education for Sustainable Development, 9(1), 4–26.

Molthan-Hill, P., Worsfold, N., Nagy, G. J., Leal Filho, W., & Mifsud, M. (2019). "Climate Change Education for Universities: A Conceptual Framework from an International Study." Journal of Cleaner Production, 226, 1092–1101.

National Research Council. (2010). Advancing the Science of Climate Change. National Research Council.

Nett, K., & Rüttinger, L. (2016). *Insurgency, Terrorism and Organised Crime in a Warming Climate. Analysing the Links Between Climate Change and Non-State Armed Groups.* Berlin: adelphi. Retrieved from https://www.adelphi.de/en/publication/insurgency-terrorism-and-organised-crime-warming-climate

Ngepah, N., & Mwiinga, C. R. (2018). "The Impact of Climate Change on Gender Inequality in the Labour Market: A Case Study of South Africa." Sustainability, 14(20), 13131.

Nolte, C. G., Dolwick, P. D., Fann, N., Horowitz, L. W., Naik, V., Pinder, R. W., & Ziska, L. H. (2018). Air Quality. In D.R. Reidmiller, C.W. Avery, D.R. Easterling, K.E. Kunkel, K.L.M. Lewis, T.K. Maycock, & B.C. Stewart (Eds.), Impacts, Risks, and Adaptation in the United States: Fourth National Climate Assessment, Volume II (pp. 512–538). U.S. Global Change Research Program.

Osama, T., Brindley, D., Majeed, A., Murray, K. A., Shah, H., Toumazos, M., & Meinert, E. (2018). "Teaching the Relationship Between Health and Climate Change: A Systematic Scoping Review Protocol." BMJ Open, 8(5), e020330.

Rumble, O., & Gilder, A. (2022). "Climate Change Mitigation Law and Policy in Africa." In O. Rumble & A. Gilder (Eds.), Research Handbook on Climate Change Mitigation Law (pp. 240–261). Cheltenham, UK: Edward Elgar Publishing.

Sachs, J., Lafortune, G., Kroll, C., Fuller, G., & Woelm, F. (2022). "From Crisis to Sustainable Development: The SDGs as Roadmap to 2030 and Beyond." Sustainable Development Report 2022. Cambridge University Press. https://doi.org/10.1017/9781009210058

Sathaye, J., et al. (2011). Estimating Risk to California Energy Infrastructure from Projected Climate Change. CEC Publication CEC500-2012-057.

Schmidt, S., Neumann, B., Warwern, Y., Durussel, C., Unger, S., & Visbeck, M. (2017). "SDG 14 Conserve and Sustainably Use the Oceans, Seas and Marine Resources for Sustainable Development." In D. J. Griggs, M. Nilsson, A. Stevance, & D. McCollum (Eds.), A Guide to SDG Interactions: From Science to Implementation. International Council for Science.

Selby, D. (2007). "As the Heating Happens: Education for Sustainable Development or Education for Sustainable Contraction?" International Journal of Innovation and Sustainable Development, 2(3–4), 249–267.

Sibanda, A., & Manik, S. (2023). "Reflecting on Climate Change Education (CCE) Initiatives for Mitigation and Adaptation in South Africa." Environmental Education Research, 29(12), 1814–1831.

Siegner, A., & Stapert, N. (2020). "Climate Change Education in the Humanities Classroom: A Case Study of the Lowell School Curriculum Pilot." Environmental Education Research, 26(4), 511–531.

Simpson, N. P., Andrews, T. M., Krönke, M., Lennard, C., Odoulami, R. C., Ouweneel, B., & Trisos, C. H. (2021). "Climate Change Literacy in Africa." Nature Climate Change, 11(11), 937–944.

Sowunmi, A. A. (2019). "Municipal Solid Waste Management and the Inland Water Bodies: Nigerian Perspectives." In H. E. M. Saleh (Ed.), Municipal Solid Waste Management. IntechOpen. https://doi.org/10.5772/INTECHOPEN.84921

Strozier, C. B., & Berkell, K. A. (2014). "How Climate Change Helped ISIS." HuffPost. Retrieved from https://www.huffpost.com/entry/how-climate-change-helped_b_5903170

Suter, M. (2018). "An Update of Yemen's Water Crisis and the Weaponization of Water." Atlantic Council. Retrieved from https://www.atlanticcouncil.org/blogs/menasource/an-update-on-yemen-s-water-crisis-and-the-weaponization-of-water/

Team, C. W., Pachauri, R. K., & Reisinger, A. (2007). Climate Change 2007: Synthesis Report. Contribution of Working Groups I, II and III to the Fourth Assessment Report of the Intergovernmental Panel on Climate Change. IPCC, Geneva, Switzerland.

Tosun, J., & Leininger, J. (2017). "Governing the Interlinkages Between the Sustainable Development Goals: Approaches to Attain Policy Integration." Global Challenges, 1(9), 1700036. https://doi.org/10.1002/GCH2.201700036

Tunji-Olayeni, P. F., Adegboye, F., Oluwatobi, A., Adeyemi, G., Olagunju, O., Okoro, A., & Osabuohien, E. S. (2021, March). "Accelerating Progress on Sustainable Development Goals: Assessing Secondary School Students' Knowledge of Climate Change Actions." IOP Conference Series: Earth and Environmental Science, 665(1), 012041.

Ugwuede, C., & Dike, C. F. (2021). "Climate Change and Terrorism in Northern Nigeria." In U. Tar et al. (Eds.), Terrorism and Counter Terrorism in Nigeria: The Domestic and Foreign Policy Dimensions (pp. 257–278). Nigerian Defense Academy Press and Nigerian Institute of Foreign Affairs.

UNCC (United Nations Climate Change). (2022). Climate Change Education. Retrieved from https://unfccc.int

UNDP. (2017). Zimbabwe Human Development Report 2017 Climate Change and Human Development: Towards Building a Climate Resilient Nation. UNDP Briefs. Retrieved from https://www.undp.org/sites/g/files/zskgke326/files/migration/zw/UNDP_ZW_2017ZHDR_Briefs---Climate-Change-and-Education.pdf

UN-DESA. (2018). Advancing the 2030 Agenda: Interlinkages and Common Themes at the HLPF 2018. Retrieved from https://sustainabledevelopment.un.org/content/documents/18777Interlinkages_EGM_Summary_Report_2018.pdf

UNEP. (2021). UNEP Food Waste Index Report 2021. Retrieved from https://www.unep.org/resources/report/unep-food-waste-index-report-2021

UNICEF. (2019). It Is Getting Hot: Call for Education Systems to Respond to the Climate Crisis. UNICEF East Asia and Pacific Regional Office.

United Nations. (2022a). All About the NDCs. United Nations Climate Action. Retrieved from https://www.un.org/en/climatechange/all-about-ndcs

United Nations. (2022b). Water and Climate Change. Retrieved from https://www.unwater.org/water-facts/water-and-climate

-change#: ~:text=Flooding%20and%20rising%20sea%20levels ,toilets%20and%20wastewater%20treatment%

United Nations. (2023). Climate action. Retrieved from https://www.un.org/en/climatechange/climate-solutions/education-key-addressing-climate-change

U.S. Global Change Research Program. (2014). "Human Health." In J.M. Melillo, T. (T.C.) Richmond, & G.W. Yohe (Eds.), Climate Change Impacts in the United States: The Third National Climate Assessment (pp. 220–256). Retrieved from https://doi.org/10.7930/J0PN93H5

USAID. (2022). Global Health: Nigeria. Retrieved from https://www.usaid.gov/nigeria/global-health

Vaughter, P. (2016). Climate Change Education: From Critical Thinking to Critical Action. Policy briefs no 4. United Nations University Institute for the Advanced Study of Sustainability (UNU-IAS), Tokyo, Japan.

Wadanambi, R. T., Wandana, L. S., Chathumini, K. K. G. L., Dassanayake, N. P., Preethika, D. D. P., & Arachchig, U. S. P. R. (2023). "The Effects of Industrialization on Climate Change." Journal of Research Technology and Engineering, 1(4), October 2020.

World Bank. (2022a). Employment in Agriculture (% of Total Employment) (Modeled ILO Estimate)—Nigeria. Retrieved from https://data.worldbank.org/indicator/SL.AGR.EMPL.ZS?locations=NG

World Bank. (2022b). What you Need to Know About Food Security and Climate Change. Retrieved from https://www.worldbank.org/

Conclusion

Elena V. Shabliy

Is Time renewable? It is a rhetorical question, but we need to act immediately to prevent negative consequences of climate change. While climate change may not yet have reached the level of apocalyptic discourse, given the possibility that humanity may find a way to develop new technologies or relocate, as some scholars suggest, it should still be seriously embraced and considered. Along the way, many communities may suffer from neglect and misinformation. This book demonstrates that smart technological advancements, as well as investment strategies, may be extremely important in combatting the climate change problem. A positive attitude and constant work in this direction of the leading economies will alter the landscape of the current neglect. The climate has been changing for thousands of years, but this time it is a human-induced problem. Future generations will benefit from the current discourse and question the neglect of the planet. It is not too late to address this *anthropogenic* problem. The development of AI could be helpful as well in addressing all these urgent issues. It is the *responsibility* of our generation to make it right, despite the current ongoing world conflicts.

A gradual change in people's minds will guarantee an ultimate success. The lives of millions of people depend on our action and pro-active behavior as well as change of thinking at this juncture. With many countries participating and recognizing this problem, a positive outcome is inevitable. More and more countries, governments, businesses, leaders, scholars, and scientists choose the environment-oriented and climate caring approach. This book is structured as follows: the discussion of Critical Raw Materials (CRMs) and their profound influence on renewable energy development opens up this book. This analysis considers technological, supply, and environmental risks, identifying investment approaches along with economic and financial policy design aimed at mitigating climate change. Chapter 1 also emphasizes the need for a pro-active action and an environmental policy advancement strategy to better tackle the problem of climate change and ensure a smooth clean energy transition pathway. Chapter 2 concentrates on the Swedish steel industry. Skoglund argues that energy transitions have impacted steel production processes. In Chapter 3, Gordon discusses International Environmental Justice as a human rights conundrum. This chapter advances the international environmental justice dialogue by, as Gordon argues, bringing light to *best available* and *best control technology* that could be useful assessing approaches in creating positive impact for mitigation and adaptation to international environmental justice problems. Chapter 4 concentrates on observations of the Michigan Environment, Great Lakes, and Energy (EGLE) Council on Climate Solutions. In this chapter, Lorr elaborates on the limits of (neo)/liberal process and how energy justice and other "justice" forward perspectives may be informing these processes to what end.

Chapter 5 provides an analysis of the sustainable electric energy devotement in Brazil, observing recent trends in wind energy development, and these trends are active. Chapter 6 demonstrates how the hierarchical framework can decentralize the distribution systems' market operations using a bottom-up perspective while addressing the challenges of scalability and interoperability. In Chapter 7, the discussion centres on the renewable energy transition in the Global South. Gz. MeeNilankco Theiventhran argues that energy transitions heavily focus on renewable technologies, where innovations will create pathways for the widespread diffusion of renewables even in unexpected places. In the global South, innovation and political economy have influenced policymaking in the energy sector over the past decade. Chapter 8 presents a case study analysis of India's multilateralism in renewable energy, using an inductive method; it is qualitative research involving a political and economic approach. Secondary data has been obtained through a thorough qualitative analysis of government reports, articles, books, research papers, journal chapters, websites, and newspapers. Official reports uploaded to government sites have been studied. These methods allow a better understanding of the relationship between India and the globe in renewable energy diplomacy in the twenty-first century, in general, and in the post-pandemic world, in particular. Chapter 9 underscores the significance of climate change education as an important strategy for human capital and investment. These chapters show that the development of green innovations, technology, and effective policy design help in the *best-available* climate change adaptation and mitigation strategies. It is important to invest in low-carbon technologies and cut investments in high-emitting sectors, and decrease the use of fossil fuels. The climate change dialogue is global, with country parties showcasing their persistent interest in solving the climate change problem. The new Business Roundtable paradigm presents a departure from the more traditional approach, which focuses on shareholder primacy, meaning that it moves away from immediate gains rather embracing sustainable practices. In the traditional approach, there is a focus on shareholder primacy referring to the idea that a company's primary responsibility is to maximize value and profit only for its shareholders. This approach focused on profit maximization for its shareholders, shareholders returns, and stock prices. At the same time, in the 2019 Business Roundtable, a new approach is suggested, where interests of all *stakeholders* should be considered, as in the more traditional corporate focus on shareholder primacy. This new approach is stakeholder-centric and stakeholder-inclusive. This new approach is revolutionary in sense of new sustainable businesses developments.

Index

About the Contributors

Edidiong Samuel Akpabio is a lecturer in the Department of Political Science at Trinity University, located in Yaba, Nigeria. His research interests encompass food and water security, social problems in Africa, and numerous other sub-themes in African development. These interests were cultivated during his postgraduate studies at the University of Benin and have since been further developed, largely due to his current role teaching courses in African Politics. He is the author of the work "Attaining SDGs in Africa through Bilateral and Multilateral Partnerships" and has recently co-authored "Food Security and Social Protection in Africa: Re-establishing the Nexus." This work appeared in one of the largest book projects concentrating on the UN Sustainable Development Goals. Edidiong Samuel Akpabio's contributions through his research, teaching, and publications are significant in advancing the understanding and addressing the challenges of African development. He can be reached on edidiong.akpabio@trinityuniversity.edu.ng & esakpabio@yahoo.co.uk.

Sedat Alataş is an Assistant Professor of Economics at Aydın Adnan Menderes University (ADU) in Türkiye. He holds a PhD in Economics. He was previously a postdoctoral research fellow in the Laboratory for Climate Change Economics at the Higher School of Economics (HSE) University in Russia. He participated in the Advanced Studies Program with a scholarship organized by the Kiel Institute for the World Economy in Kiel, Germany, prior to finishing his doctoral thesis. His current research mainly focuses on climate change, energy and environmental economics, material economics, resource efficiency, economic growth, sustainable development, and applied panel data econometrics. He published several research papers on these topics. He also teaches a variety of undergraduate and graduate courses at ADU and HSE University, including Economic Growth, Econometrics, and Microeconomics.

Aditi Basu completed her master's in Political Science from Jamshedpur Women's College, India. Her research interests include Indian Foreign Policy, International Relations, Feminist Power Politics, Soft Power Diplomacy and Climate Diplomacy.

Dr. Christine Chen is an Associate Professor with the Department of Electrical and Computer Engineering, The University of British Columbia, Vancouver, BC, Canada, where she is affiliated with the Electric Power and Energy Systems Group. Her research interests include power system analysis, control, and operation. Prior to joining UBC, she received the BASc degree in engineering science from the University of Toronto, Toronto, ON, Canada, and the MS and PhD degrees in electrical engineering from the University of Illinois at Urbana-Champaign, Urbana, IL, USA. She is a recipient of a Best Paper Award from the IEEE Transactions on Energy Conversion in 2018 and a "Best of the Best" Prize Conference Paper Award at the Power and Energy Society General Meeting in 2022. She also co-authored the 2021 IEEE Power and Energy Society Prize Paper. Christine has served on the editorial boards of the IEEE Transactions on Power Systems, IEEE Transactions on Energy Conversion, and IEEE PES Letters.

Chinonye Frances Kamsi Dike is a graduate of zoology from the University of Ibadan, Nigeria. She is a researcher with an insatiable quest for issues of sustainability as gleaned from her research articles in high impact publications and Scopus indexed conference proceedings. She can be reached at nonyedike1@gmail.com.

Kimarie Engerman, PhD, is a full Professor of Psychology, and Dean of the College of Liberal Arts and Social Sciences at the University of the Virgin Islands.

Frederick D. Gordon is an Associate Professor and MPA Director for East Tennessee State University. He is the author of several publications, including *Freshwater Resources and Interstate Cooperation Strategies to Mitigate an Environmental Risk* (2008), lead editor for *International Environmental Justice Competing Claims and Perspectives* (2012). His latest book is *The Damming of the Presidency: How Environmental Factors Impact a Political Campaign* (2021). He received his PhD in Political Science from the University of Southern California and master's in Public and International Affairs from the University of Pittsburgh.

Dr. Burcu Hiçyılmaz is an Assistant Professor at Aydın Adnan Menderes University in Türkiye. Dr. Hiçyılmaz earned her bachelor's degree in

Economics from Gazi University between 2009 and 2013. In 2015 and 2019, she obtained her master's and doctoral degrees from the Aydın Adnan Menderes University Social Sciences Institute Economics Programme. Her primary research interests include climate change, material and energy economics.

Etem Karakaya is a Professor at Eskişehir Osmangazi University in Türkiye. Previously he worked in different universities as a Professor in Türkiye as well as a visiting researcher at Rutgers University in New Jersey, USA. Between 2005 and 2008, he worked as a project manager and Türkiye desk officer in the "climate change and energy" unit of the European Environment Agency, the EU institution headquartered in Copenhagen. While working for the European Environment Agency, he served as a European delegation to the European Commission, UNFCCC, OECD, and other institutions. Dr. Etem Karakaya has been working on climate change economics and international dimensions, carbon markets, resource efficiency, European economy, and international trade. He has published several research papers and reports on these topics.

Elizabeth A. Keenan is an assistant professor of business administration in the Marketing Unit. She teaches the Purpose-Driven Marketing course in the MBA elective curriculum (EC) and previously taught the Marketing course in the MBA required curriculum (RC). She is affiliated with the Behavioral Insights Group at the Harvard Kennedy School's Center for Public Leadership as well as with the Harvard University Center for the Environment. Professor Keenan's research explores individuals' prosocial choices and behaviors within the domains of charitable giving and environmental sustainability. Her research has been published in Science, the Journal of Consumer Research, and Nature Climate Change, and it has been cited by media outlets, including NPR, The Huffington Post, The Wall Street Journal, and U.S. News & World Report. Professor Keenan earned her PhD in marketing at the Rady School of Management at the University of California, San Diego; an MAS in marine biodiversity and conservation, also at UC San Diego; and a BS in biology at Loyola Marymount University. Prior to her doctoral studies, Professor Keenan spent ten years in nonprofit management and education at the Aquarium of the Pacific.

Roohallah Khatami received his BSc, MS, and PhD degrees respectively from Iran University of Science and Technology, Tehran, Iran, in 2007, Amirkabir University of Technology, Tehran, Iran, in 2013, and the University of Utah, Salt Lake City, UT, USA, in 2019, all in electrical engineering. He was a Postdoctoral Research Associate with the Department of Industrial &

Systems Engineering, Texas A&M University, College Station, TX, USA, a Postdoctoral Research Fellow with the Department of Electrical and Computer Engineering, the University of British Columbia, Vancouver, BC, Canada, and he is currently an Assistant Professor in the School of Electrical, Computer, and Biomedical Engineering at Southern Illinois University at Carbondale, IL, USA. His research interests include power systems operation and electricity markets.

Dmitry Kurochkin is a Senior Research Analyst at Harvard University. He graduated *magna cum laude* from Lomonosov Moscow State University where he was majoring in Physics. Kurochkin earned his PhD in Mathematics at Tulane University and holds master's degrees in a variety of fields, including Chemistry, Statistics, and Economic Analysis and Policy.

Michael J. Lorr is Professor of Sociology and Director of the Community Leadership Program at Aquinas College in Grand Rapids, MI. His work on urban, environmental and cultural sociology has recently appeared in the *Journal of Contemporary Ethnography*, *Nature & Culture*, *Humanity and Society*, *Journal of Youth and Adolescence*, *New Directions for Student Leadership* and as book chapters in a number of edited volumes such as *Neoliberal Chicago*, *The Greening of Everyday Life*, and *Skateboarding: Subculture, Sites, and Shifts*.

Sakshi Mishra is Senior AI Engineer in the Autonomous Systems group at Microsoft, Redmond, WA, USA working on Project AirSim—Aerial Autonomy. Prior to joining Microsoft, Sakshi was leading REopt Lite API and its open-source effort at the National Renewable Energy Laboratory (NREL), Denver, CO, USA as a research engineer. Sakshi also led the deep-learning learning based Predictive Analytics work at NREL. She received her bachelor's degree in electrical and electronics engineering from VIT University, India, and her master's degree in EST&P from Carnegie Mellon University, as an academic excellence awardee. Sakshi is an IEEE Senior Member and holds a Professional Engineer License in the state of California.

Lucigleide Nery Nascimento holds a B.S. in Economics from the Universidade Católica do Salvador, Brazil. She earned an M.S. in Natural Resources—Environmental Conservation and a Ph.D. in Natural Resources and Environmental Studies from the University of New Hampshire, United States. She is a Specialist in Production of Economic, Social and Geoenvironmental information (Especialista em Produção de Informações Econômicas, Sociais e Geoambientais) at the Superintendence of Economic and Social Studies of Bahia (Superintendência de Estudos Econômicos e

Sociais da Bahia)—SEI, Salvador, Bahia, Brazil. The views and opinions expressed in the chapter are those of the author and do not necessarily reflect the official policy or position of any agency of the Bahian government.

Ogunbadeniyi Adeola Oluwakemi is a lecturer in the department of Political Science University of Benin, Benin City, Nigeria. She has over a decade of teaching experience in the tertiary education sector. The author can be contacted via adeola.ogunbadeniyi@uniben.edu.

Elena V. Shabliy received her Interdisciplinary PhD from Tulane University in 2016. In 2009, she earned a Master of Liberal Arts degree from Tulane. She was a Visiting Scholar at Harvard University, Columbia University, Boston University and NYU. In 2018, she was a Postdoctoral Fellow at Harvard University.

Annika Skoglund is Associate professor at Uppsala University, Sweden, and Senior Lecturer at the University of Bristol Business School, UK. Embedded in European corporations, Skoglund ethnographically researches decarbonization technology and transformations towards sustainability. She has co-authored a research monograph titled *Climate Activism—How Communities Take Renewable Energy Actions Across Business and Society* (2022), and her research appears in journals such as *Organization Studies*, *Human Relations*, *Critical Policy Studies*, *Sustainable Development* and *The Sociological Review*.

Gz. MeeNilankco Theiventhran holds a PhD in Human Geography from the University of Oslo. His research mainly concerns equity and justice aspects of energy transitions in the Global South and pays specific attention to geopolitical dynamics and policy challenges confronting developing countries in achieving clean energy transitions. He is the course leader of the master's programme in International Development Studies at the International Summer School, University of Oslo and a visiting lecturer at the sociology and human geography department, University of Oslo.

Ifeanyi E. Uzochukwu is a Lecturer and Faculty Member at the Faculty of Agriculture, University of Nigeria, Nsukka (UNN), where he is actively engaged in teaching, research, and consultancy. With over 12 years of experience, Ifeanyi's activities focus on optimizing reproduction, production, and sustainability in animal and aquaculture systems. Currently, Ifeanyi is pursuing a PhD in Environmental Sciences at the University of Debrecen, Hungary. His research concentrates on the bioaccumulation and toxic effects of anthropogenic contaminants in freshwater organisms, reflecting his passion for environmental health and animal product safety. He can be reached at ifeanyi.uzochukwu@unn.edu.ng.

www.ingramcontent.com/pod-product-compliance
Ingram Content Group UK Ltd.
Pitfield, Milton Keynes, MK11 3LW, UK
UKHW041835031026
15587UKWH00002B/782

9 781666 965810